Propagate Your Own Plants

Propagate Your Own Plants

Wilma Roberts James
Arla Lippsmeyer

NATUREGRAPH PUBLISHERS, INC.

Library of Congress Cataloging in Publication Data

James, Wilma Roberts, 1905 -
    Propagate your own plants.

    Bibliography: p.
    Includes index.
        1. Plant propagation.   I. Lippsmeyer, Arla,
joint author.   II. Title.
SB119.J35   1978   635'.04'3   78-18248

ISBN 0-87961-073-5 Cloth Edition
ISBN 0-87961-072-7 Paper Edition

Naturegraph Publishers, Inc., Happy Camp, CA 96039

*To Sevrin Housen for his steadfast encouragement
and editorial skill, I am everlastingly grateful.*

ACKNOWLEDGMENTS

I wish to express special gratitude to Arla Lippsmeyer for her accurate and detailed line drawings which so admirably complement the written text. For offering advice and criticism based on their own expert knowledge of plants, "thanks" go to Barbara and Vinson Brown. I am also grateful to the librarians of Sacramento City and County libraries for their helpfulness and skill. Appreciation is expressed to Rick Hill for the cover picture of cherry laurel he designed from a photograph by Forest Bliss. Finally, I must acknowledge my indebtedness to all those not mentioned here who helped make the writing of this book possible.

Wilma James
Sacramento, California

TABLE OF CONTENTS

INTRODUCTION

This book is designed for new or advanced gardeners who want to multiply their plants at the lowest possible cost. It guides the reader in selecting and propagating plants, including sturdy American natives, all-time favorites and useful herbs. Emphasis is placed on those species and varieties which can be easily reproduced from their own vegetation. Vegetative reproduction provides quicker results than seed germination and is also the sole means of reproducing some species. By this method economy-minded gardeners can duplicate plants without expense, and every gardener can be assured each offspring will be identical to its parent.

Throughout the main text, requirements and methods for propagation follow individual plant descriptions and illustrations. Included is information on how to compost, cultivate, control pests, as well as ways to utilize herbs and other health-promoting plants. Although remedies for healing are described, none of them should be interpreted as prescribing. Medical plants should only be used upon consultation with and the consent of a physician.

A goodly portion of the plants discussed in detail come as natives from all regions of the United States. Early settlers from Europe brought others which escaped cultivation. These vigorous plants all adapt well to garden situations, being attractive and remarkably carefree. The information given on each plant's origin and growth habits will aid growers in determining which candidates are most suitable for their indoor and outdoor gardens.

The index to this book is extensive. Besides listing the specific plants covered in the text, it gives readers the opportunity to discover additional plants which respond similarly. For example, the needs of evening primroses are explained on page 84. For other plants with identical propagational requirements, the reader is referred to *see* page 84. In this way, PROPAGATE YOUR OWN PLANTS thoroughly covers the basic needs of well over seven hundred plants.

HOW TO GET STARTED

Growing your own new plants is easy and fascinating. There are a number of ways you can obtain propagation material without having to buy it. Here are some suggestions for getting started.

1. Begin at home by increasing your own plants from cuttings, divisions, root cuttings, suckers, or by layerings.

2. Be aware of what plants your friends and neighbors are growing. They may be happy to share plant parts and/or advice on care.

3. Join a garden club. You will be able to take part in exchanges and sales, and helpful information will come from within the group.

4. Watch newspapers for single-plant society sales to secure plant material that would not be available otherwise.

5. Save plants from destruction by digging them up from development sites. Once they are reestablished, propagate them.

6. Search around abandoned farms, along roadsides and fences for plant materials. You can, for example, start new rhododendrons from leaf cuttings with heels attached. A great number of perennials and many woody plants will oblige you with root cuttings, and you can take them without removing a whole plant.

7. Investigate along streams, lakesides, marshes, woodlands and open hill country for native plants. Remove only minimal portions for propagation. To preserve the environment, cuttings are recommended.

8. Parks require permits to remove plants, but they may be salvaged on public lands where vegetation is to be destroyed.

9. Since fragile native plants seldom live after removal from their habitats, it is best to collect their seeds. Do this soon after the seeds fall. Leave enough seed to guarantee a continual plant population.

10. Whatever course you pursue, you will often find it necessary to try plants or seeds from nurserymen and seedsmen. See pages 139-140, for a listing of commercial sources. Specialists in native plants can be found on these pages as well.

PART I

GENERAL PROPAGATION AND CARE OF PLANTS

CUTTINGS

Essentially, cuttings are parts of a plant which have been severed from a parent plant. The usual procedure is to use a sharp knife to cut off a branch, stem, leaf, or root, and insert it in a rooting medium. Under favorable conditions, cuttings form roots and develop into plants which are identical to their progenitors. Vegetative propagation by this method is an easy and economical means to obtain a large quantity of sizable plants more quickly than from seed.

Stem Cuttings

A great number of plants propagate easily from stem cuttings. A partial list includes geraniums, coleuses, many cacti and other succulents. These cuttings, consisting of sections or tips of the stem of a plant, should be taken from good growth. Choose tip or stem ends from the base of a plant or from its side branches. Cut approximately 4-inch lengths from the plant, not more than ½ inch below joints. Strip the lower leaves that might touch the rooting medium, but retain as much foliage as possible. Insert the cutting about one-half its length in a rooting medium. To prevent rotting of geraniums, succulents, or any species that oozes a heavy sap, wait until the cut dries and sets a callus over its wound.

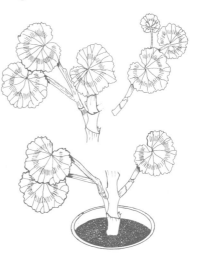

Roots are most likely to appear from the section of the stem just below the joint. After roots develop, the stem is ready for transplanting.

Hardwood Cuttings

These cuttings are the simplest type of stem cuttings. Deciduous woody plants such as dogwood, fig and grape are frequently used. For cuttings, select hard or well-ripened healthy shoots from the same season's growth, between November and February, after the leaves have fallen. Cut the sticks 6 to 12 inches long, making sure they possess two or more buds or nodes, from which shoots may develop.

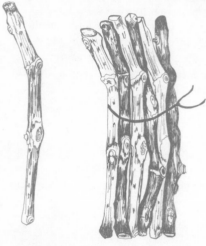

During autumn and winter months, short cuttings can be placed in the open ground or in a container. For longer cuttings, tie the sticks in small bundles with string or wire, pointing the tops in one direction. Stand the bundles inside boxes of damp sand with only one or two buds exposed, and store them in a cool, frost-free shelter such as a garage or basement. Never allow cuttings to dry out while they are forming calluses at their ends.

Come spring, root the cuttings individually in containers, or in beds 4 to 6 inches apart, buried to the uppermost pair of buds.

Semihardwood Cuttings

These cuttings are taken from broad-leaved evergreen plants such

as acacia, jasmine and laurel. They are taken in late summer when growth is firm. Cut the new side shoots 3 to 6 inches in length, directly below a node. Remove the lower leaves to reduce water loss. More than one cutting can be made from a side shoot by leaving a node with its leaves at the top of each succeeding section. Shoots will develop from the dormant buds at the top of the node. Insert semihardwood cuttings in a rooting bed filled with equal parts of sand and peat.

Softwood Cuttings

The term "softwood" applies to cuttings of soft or green wood taken from a current year's growth of deciduous or evergreen shrubs and trees. These cuttings are taken between June and September. Pieces of a vigorous branch, 2 to 6 inches long, are cut below a node, the lower branches are removed, and the cuttings planted under glass in sand, or in a humid greenhouse. When new roots form and top growth appears, the cuttings may be transplanted. Evergreens

never be allowed to wilt before it is put in a rooting medium.

Leaf Cuttings

Here is a rapid means of increasing a wide variety of plants with thick fleshy leaves. Christmas cactus, *Epiphyllum*, *Crassula* and many *Sedums* can be rooted by cutting off a leaf, or section of leaf, with or without a stalk. Place cuttings with half of their length base down in a mixture of one-half potting soil and one-half sand. Space individual leaves so they do not touch each other in the planter, keeping them rather dry and uncovered. Small roots form at the base and leaves will develop above them. For best results, maintain a temperature of 70° F.

Root Cuttings

A number of plants with true roots may be propagated by using this method. They include erect type blackberries, raspberries, wisterias, some ferns, and certain shrubs and trees which do not root readily by stem cuttings.

Dig up the roots of deciduous plants in autumn after the leaves have fallen. Potted plants can be knocked out of their containers. Choose main roots and cut them into 1- to 3-inch pieces, each containing an eye or bud growth. Place the pieces horizontally in moist sand about one-half inch below the surface and water thoroughly. Store in a cold

but frost-free location to form calluses. In the spring, plant in a horizontal position 2 inches deep in sandy soil. Shoots will form, sprouting into newly rooted plants. They are ready for transplanting when the shoots reach 3 inches.

OTHER SIMPLE METHODS OF VEGETATIVE PROPAGATION

Runners

A runner is a stem which develops near the base of a leaf at the crown of a plant and produces new plants at the joints along its length. The best example of this form of reproduction is the strawberry. For this method, cut off the runners during the growing season, lift up the new plants where they have rooted, and replant.

Another method is to fill a small pot with potting soil, putting the young plant into contact with the soil so it can form strong roots while still attached to the parent plant. Use a hairpin to hold each new plant in place until well established. Then cut the runner close to the plant.

Suckers and Offsets

A sucker crops up from the lower part of a plant stem or trunk, while an offset grows from a plant base and usually roots at the tip to form a new plant. Typical plants which produce suckers are various shrubs and trees, often suckering from the ground. Offsets are common in artichokes, houseleeks and yuccas.

Cut off the suckers and off-sets singly from the parent plant. If basal suckers have roots, plant in individual pots. Place unrooted cuttings in a sand-filled box. Keep moist and shelter from direct sunlight while roots develop.

Division

A great number of perennial plants, such as bamboo and many herbs can be propagated by division. Divide in the spring when new growth begins, or in the fall with late blooming plants. Take up the smaller plants and pull the roots apart by hand. Sever plants with thick fleshy roots, using a large kitchen knife or hatchet. Shake the soil from the roots, then cut through the section where the roots and top growth join. Slice the clump so as to provide at least one bud (eye) and some roots. Replant the divisions separately.

Divisions must be replanted immediately. Water each plant thoroughly at first, and then whenever soil surface feels dry. Protect the new plants from direct sunlight until leaves are firm.

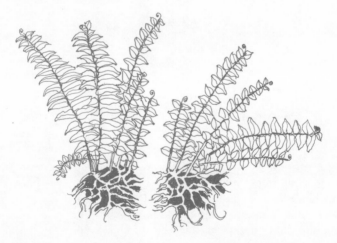

Rhizomes

Normally, a rhizome or rootstock is a thickened underground stem which grows horizontally close to the soil surface and sends out creeping roots for food gathering. The iris and brake fern are examples. A stem tuber is much the same as a rhizome. It is an underground stolon swollen with stored food reserves. A common example is the Irish potato.

Most rhizomes and stem tubers are best propagated in spring or fall. Use a sharp knife to cut them into sections, ensuring that each section has buds and is large enough to contain sufficient food. Shoots grow from the buds, with the majority of the roots forming along the base of each shoot.

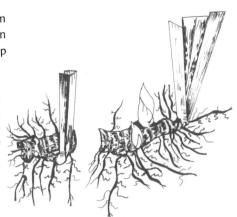

Layering

This is the rooting of a branch of a woody plant while still attached to the parent plant. It is a method employed for a wide variety of vining plants, including shrubs with stems long or flexible enough to be brought down to the ground.

For ordinary layering, which produces one plant from each layer, a strong shoot of the previous season is used in the spring. A slit, notch, or a slicing off of some of the bark to be layered is needed in forming a callus. Make a right-angle twist so the tip of the shoot stands erect. Anchor the shoot firmly in the soil with a notched peg or a piece of bent wire. Cover the layer with a mixture of sand and peat, allowing the shoot to protrude about 3 inches above the surface. Keep moist at all times.

By late fall of the same year the layered branch of most species will be rooted and should be severed from the parent plant. Transplant the reproduction in the spring.

Mound Layering

This method of layering is used primarily for low-growing shrubs with stiff upright stems which produce many shoots from a crown near

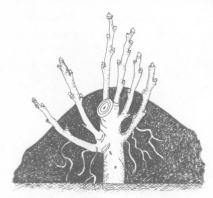

the ground. Flowering quince, black currant, and gooseberry are examples.

First cut back the plant almost to ground level in early spring. When new growth appears, half-bury the new shoots in soil. Then form a mound over the old crown. The newly developed shoots are forced to root and are removed separately a year later.

ROOTING MEDIUMS FOR CUTTINGS

Rooting mediums should be sterile—that is, contain no weed seeds, bacteria, or harmful fungi. The following materials are commonly employed.

Sand

Use coarse builder's sand. Sterilize by pouring boiling water through it. Sand is heavy and has poor moisture retention, but works well with many kinds of cuttings.

Perlite

A white, porous volcanic mineral, perlite has been expanded by heating to form very light kernels. It does not absorb moisture but is valued for its aeration capacity and light weight. Roots start to develop quickly in perlite.

Vermiculite

This is soft, brownish, heat-expanded mica. Spongelike, it holds moisture nicely and allows cuttings to make strong roots. There is one drawback to this product, however: once used, vermiculite tends to lose its water-holding ability.

Sphagnum Moss

Collected from bogs, sphagnum moss is packaged and sold in whole pieces, either fresh or dried. Highly water absorbent, shredded sphagnum is advantageous as a rooting medium and as an additive to other materials. Whole pieces are used for lining wire hanging baskets.

Peat Moss

Much sphagnum moss is sold in a fine, decomposed state known as peat. It provides acidity, aeration, and retains moisture well when combined with perlite or sand. Cuttings from acid-loving plants root excellently in peat mixes.

Leaf Mold

This soft, flaky, organic matter forms from fallen leaves that have accumulated for several years. It can be found under the top layer of leaves in the woods. Humus-bringing leaf mold combines splendidly with sand or shredded sphagnum to propagate cuttings.

Water

For generations, rooting cuttings in water has been a popular and easy method of propagating plants. Cuttings are put in a small glass container of water with one-third of the stems submerged. The water must be replenished as it evaporates. Place cuttings and all in a window facing east or west. To counteract tangled roots, add small amounts of sand every day or two. The sand will also help tender roots to adjust to potting soil at transplanting time.

UNITS FOR ROOTING CUTTINGS

Double Pots

An excellent method for rooting stem cuttings not in need of high humidity is the use of double pots. Coleus, geranium and honeysuckle cuttings fall into this group.

First, plug the hole at the bottom of a small clay pot with a cork or chewing gum. Insert it into the middle of a shallow 6-inch clay pot filled with premoistened rooting medium such as sand or vermiculite. Fill the small pot with water and never allow it to dry out. Seepage from the small pot keeps the rooting medium supplied with water. Punch holes in the ring of rooting medium approximately three-fourths of an inch deep. Insert cuttings upright, making certain that the base of each stem touches the bottom of the hole. Then press the rooting medium firmly down around the stems.

Place double pots in bright light, but not directly in the sun; maintain the temperature at about 75° F. To prevent wilting, spray with a fine mist. Check for roots in a week. Plants with roots 1 inch long are ready for transplanting. Put back those with shorter roots and wait awhile longer.

Single Pot Method

A 4-inch pot is adequate to propagate three or four cuttings. Fill the pot with thoroughly soaked vermiculite or perlite, and drain off excess water. For softwood cuttings, use a moist mixture of sand and peat. Stand the cuttings upright in the rooting medium, pressing the medium around them. If the cuttings require high humidity, cover them with a piece of glass, or a plastic bag. Place the pot in a bright, but not sunny, location and wait for 1-inch roots to develop.

To root a single cutting that requires high humidity, cover the cutting with an inverted glass jar. If much moisture collects inside the glass, lift the covering for an hour or two to dry out.

A single cutting not requiring high humidity may be wrapped with a dampened rooting medium at its base. Take a 6-inch square of plastic wrap, tie it securely around the rooting medium with a twistwire, and additional water will not be necessary for weeks.

Propagation Boxes

A plastic shoe box or similar transparent container will keep a large number of softwood cuttings, or cuttings from stems and leaves,

at 100 percent humidity. Pour 2 inches of vermiculite or perlite rooting medium into the box. Sprinkle with water until it collects in a small puddle at the bottom. Tip the box back-and-forth until you are sure the medium is completely wet. Then drain off excess water.

Insert as many cuttings as you wish, provided they do not touch each other. Put the lid on the box and place where there is bright light—but not in the sun. Maintain a temperature of about 75° F. When moisture accumulates in the box, remove the lid for awhile. The new plants should be ready to pot in about five days.

Before removing the plants from their intensive care unit, give them a naturalizing period. Open the box one-fourth inch the first day, one-half inch the second day, and a whole inch the third day. On the fifth day, take the plants out and pot.

For semihardwood cuttings use the same type of propagation box without the lid. Place these cuttings in a moistened rooting medium, shade them from direct sunlight, and provide humidity by misting. When new leaves appear, give more light. Pull out the cuttings in a few weeks to see if they are well-rooted. Pot those that are. Otherwise, replace the cuttings and wait awhile longer. Rooting time varies, depending upon the species involved.

Glass Enclosures

To root cuttings requiring high humidity, a terrarium, an aquarium, or a large glass jar is practical. The procedure is identical to that of a covered propagation box, except the top is covered with a piece of glass or aluminum foil to retain moisture. When supplied with a rooting medium, cuttings flourish inside these miniature greenhouses.

Wood and Plastic Mini-greenhouses

A wooden box 3 or 4 inches deep with holes bored in the bottom makes an inexpensive propagation box. To keep the cuttings warm and moist, bend two or more metal coat hangers over the box. Stretch clear freezerwrap over the coat hangers, or slide the box into a large, clear plastic bag and close the mouth with a twistwire. This portable enclosure offers space for rooting almost any cutting requiring humid conditions.

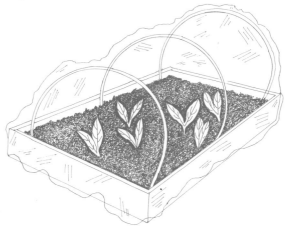

Commercial Propagation Boxes

Propagation boxes made especially for rooting cuttings are sold commercially. These indoor greenhouses come equipped with or without fluorescent lighting. One economical, all-plastic model without lightings contains forty individual compartments for cuttings. It also has a lid and self-watering tray, the water drawn upward by a wick.

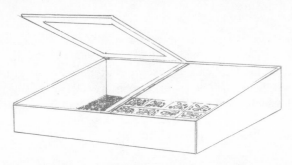

Models furnished with fluorescent illumination encourage cuttings to root in the darkest areas of rooms, halls, or basements. They are particularly advantageous for quicker rooting and can be used all year. A pair of cool-white 40-watt fluorescent tubes can provide adequate light for a separate growing area independent of outside light. Tubes should be 3 or 4 inches above the cuttings, and the hood raised as the plants grow. A light period of 10 to 12 hours produces good results. If a timer is used, it will automatically switch the lights on and off.

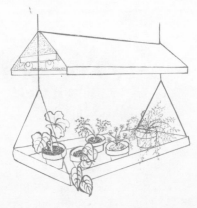

Cuttings placed under fluorescent lights require added humidity. Hand-misting should be applied twice a day. Another necessity is a humidifying tray. Layer a shallow pan or tray with pebbles or gravel. Keep the tray filled with water to the top of the pebbles. Arrange the propagating unit on top of the pebbles and water, then place under the fluorescent lights.

Low-powered heating cables are a must when cuttings are being propagated that require bottom heat. They can be bought, or made at home. The electric coils produce a steady heat night and day in a hotbed frame. Directions for making and for the proper use of them are found in many books on the subject.

Peat Pots

These pots made of pressed peat moss are for individual cuttings. Found in garden supply stores and through catalogs, they come in round or square shapes. A large number of cuttings inserted in peat pots can be placed in one propagation box, the rims touching to support each other.

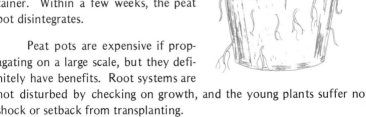

Fill each peat pot to the top with a slightly moist mixture of sand and peat, then insert a cutting. Keep the pot and rooting medium moist at all times, avoiding direct sunlight. When roots show through the pot, transplant into the garden or the next sized container. Within a few weeks, the peat pot disintegrates.

Peat pots are expensive if propagating on a large scale, but they definitely have benefits. Root systems are not disturbed by checking on growth, and the young plants suffer no shock or setback from transplanting.

Jiffy-7 Pellets

These compressed peat disks come wrapped in plastic netting. Each disk expands to seven times its original height after soaking in water, thus the name. Look for them in garden supply stores or in mail-order catalogs.

Soak a pellet in water for 10 minutes, or until totally expanded. Pierce a hole with a pencil in the open top, insert a cutting and squeeze out excess water as the pellet is pressed around the cutting. Put in a plastic bag, blow it up, and seal off the top with a twistwire. As many as three expanded pellets can stand in one bag.

Unless humidity is extremely low, the plastic bag need not be sealed. Keep the cutting moist and avoid direct sunlight. When roots appear through the netting, move the device into a 3-inch pot filled with rooting medium. As with peat pots, the root system remains undisturbed during transplanting.

The expanded pellets with cuttings may be arranged in small clay pots and stood on a humidity tray layered with pebbles. Next, add water to a level just below the top of the stones. Place under fluorescent lighting, or where there is indirect sunlight, until roots show through the netting.

POTTING ROOTED CUTTINGS

Soil

Never use soil straight from the garden. It contains weed seeds, soil insects and fungus diseases which often harm or kill potted plants. A light, loose, fast-draining soil is required. You can buy suitable ready to use soil substitutes at garden supply centers, or you may prefer to make your own.

Ready-prepared mixes are free of weed seeds and plant disease organisms. They retain moisture and plant nutrients well, and are light-weight and portable. Because ready-prepared mixes tend to compact when watered, preventing air circulation to tender roots, it is a good idea to add one part perlite to two parts of the mix. As an alternative, the same ratio of sand may be added.

Here is one formula for preparing your own soil substitute. Mix equal parts by volume of damp shredded sphagnum moss with perlite or vermiculite. To each two gallons, mix in one tablespoon of 5-10-5 fertilizer and one tablespoon of ground limestone. When potting acid-loving plants like azaleas, omit the limestone. If cacti, add two parts of sterilized sand. Mix all ingredients thoroughly.

Containers

No matter how small your space, some kind of container will sustain your young plants that are ready for transplanting. Almost any plant will grow in a container, be it a tree, shrub, or herb. They can be placed on a paved patio or terrace, on a porch, steps, balcony, even the apartment house roof.

Fruit trees and other flowering trees as well as large shrubs furnish shelter from the sun and break the wind. They demand containers large enough to hold the plant when fully grown. Wooden vegetable and fruit shipping crates collected from supermarkets or a municipal dump may be used. To avoid the risk of overwatering, bore a few drainage holes in the bottoms or about an inch above the bottoms along the sides. You may want to dismantle the crates and use the salvaged lumber to build containers of your own design and size. These planters last for years if painted with a wood preservative. To further assist drainage, put one-half inch of coarse gravel or crushed rock at the bottom of the container. Overly large or heavy containers can be mounted on casters for easy moving.

When support is needed for a young tree, shrub or vine, tie the main stalk onto a single stake with sufficient height to compensate for growth. For vining plants with clinging tendrils, a flaring trellis works well. Guard rolls are likewise available in garden supply stores. One roll makes six (18-inch diameter by 48-inch high) cages, which are good for climbing plants. The guard eliminates tying to stakes and helps prevent ground rot.

Herbs and small foliage plants grow fine in pots, cut-down half-gallon milk cartons, and in cans provided with drainage holes. A drip saucer should be placed beneath containers indoors to catch the normal runoff from waterings. In the garden, set pots of herbs among other plants to discourage harmful insects. Wood planter boxes, coming in different sizes, will hold a number of small plants. But with a saw, hammer and nails, shipping crates can be constructed into desirable shapes and sizes.

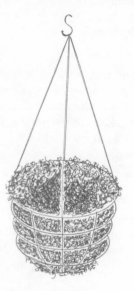

Hanging planters lined with sphagnum moss and filled with potting soil are ideal for conserving space. Many low-growing, wide-spreading plants classed as groundcovers will drape down to form lovely hanging bouquets.

Made of unglazed clay, certain containers devised for strawberries have small cup-shaped openings around the sides and wide openings at the top. These attractive containers are also excellent for growing herbs. A homemade version can be constructed by drilling large holes in the sides of a barrel.

The beauty of common vegetable plants is often overlooked. This fact can be emphasized by improvising and planting them inside of old tires or paint buckets. A 40-gallon galvanized metal trash container has the capacity to hold enough soil to sustain globe artichoke or rhubarb. Don't neglect to punch drainage holes.

Soil always dries faster in a container than in a garden bed and should be checked daily in hot weather. If the top inch feels dry, apply water until it runs out drainage holes. Excess water must be removed from a saucer within two hours after watering because standing water can damage a plant. Try to avoid splashing water on leaves and flowers when watering late in the evening. Keeping plants wet overnight encourages diseases.

Transplanting

When transplanting, remember to always start with a clean container. Soak a used clay pot in a pan of boiling detergent water for several hours to loosen remnants of any soil, salt, or algae. Then scrub it with a wire brush and scouring powder, rinse, and dry in the sun.

Even new clay pots need to be soaked in water for an hour and let dry awhile. This prevents the porous clay from robbing moisture from the soil and the plant.

Use a pot ½ to 1 inch larger than the plant's previous container. Line the bottom of the pot with a layer of coarse gravel to keep the soil mix from being washed out through the drain holes. After filling the pot part way, center the plant so its root crown is one-half inch below the pot rim. Spread the mix in and around the plant. Shake the pot back-and-forth to settle the mix. Keep adding potting soil until it is on the same level with the root's crown, then gently smooth the surface soil and thoroughly water the plant.

Place pot and all in a cool, sunless location for a week to let the plant adjust. If it wilts, do not rewater. Try misting, or seal pot and plant in a plastic bag.

Transplant whenever a pot is filled with roots. The most visible signs indicating a plant's need for transplanting are roots pushing through drain holes or showing above the topsoil. Move most plants to the next sized pot. With fast-growing plants you may want to skip a size. Several small plants may be transplanted into a large container.

BASIC PLANT CARE

Since plants can be killed by too much loving care, it is important to know something about different plants' needs. Many plants struggle along in an unfavorable environment for awhile, whereas, put in a compatible place they would thrive.

Light

Nearly all plants, especially those that flower, require some sunlight. The advantage of potted plants is that the amount of sunlight can be readily controlled by shifting their position. Remember, sunshine strikes at a different angle in the winter than in summer months. Where there is plenty of sun one season, it may be shady another. The majority of outdoor plants must have five to six hours of sunshine daily. Put sunlovers, such as vegetables, where they can receive their full quota.

Water

Overwatering or letting a plant stand in water too long causes yellowed leaves, brown tips on leaf edges, and leaf drop. It may also cause roots to rot. Watering needs depend not only on the plant species, but on the temperature, air circulation, and the type of pot. A plant growing in a clay pot requires more water than those grown in other containers. Immerse the plant and soil in a pail of cool water for about an hour. Then remove it from the water, let drain, and replace to the growing area. In general, most plants should be watered thoroughly after they become almost dry.

Many plants, especially foliage plants, like to have their leaves moistened regularly during hot weather. A temperature of around 72° F. is perfect. To create a pleasantly moist and cool atmosphere, use a nozzle to make a fine mist. This is best done in the early morning; but if plants start to droop, apply a spray of water immediately. Plants such as coleuses soon perk up once moisture is applied to the dry air around them. Most houseplants do not grow well in rooms with low humidity. Lack of humidity shows up when leaves are pale and color-less. Even though your plants are flourishing remember a fair amount of moisture in the air is conducive to good plant growth.

Fertilizer

Various kinds of barnyard manures stimulate plant growth by providing organic matter known as "humus" to the soil. This is good plant food and, if aged for a year or more, will not burn tender plants. Well-rotted animal fertilizers can be worked into the soil at planting time and applied one to three times during the growing season when plants show need of it. Scrape it into the soil surface and then water thoroughly. A level tablespoonful added to a 6-inch pot produces good results. Manure steeped in water is also an excellent soil supplement, particularly when bone meal has been mixed in the potting soil. Due to its strength, use poultry manure at half the ratio of cattle or sheep manure.

Compost is one of the best possible fertilizers to use. Make a cylinder of concrete reinforcing wire, or employ a garbage can with holes punched in the lid to start a compost pile. Heap all available grass, leaves, plant tops, discarded vegetables and kitchen scraps in a 6-inch-thick layer. Spread over this an inch layer of barnyard manure. Keep adding layers, and water to keep it damp. During warm weather, the mixture will shrink to half its size within three months. Turn the material thoroughly. Usually in another two months, a crumbly, moist, rich earth is ready to scrape around flower borders, shrubs, and trees.

Cottonseed meal is a complete, nonburning, vegetable-meal fer-tilizer that is available commercially. The preferred time for applying it is in the spring. One teaspoonful is sufficient for a 6-inch pot. For growing plants, work cottonseed meal lightly into the soil. The acidity of this meal makes it exceptionally effective on azaleas, rhododendrons, and other acid-loving plants.

Most people elect to use commercially mixed fertilizers indoors. They are sold in dry or liquid forms. Numbers on the bottles or bags refer to the proportions of the three main soil elements essential to healthy plant growth. If a label reads 5-10-5, then 5 percent of the contents is nitrogen, 10 percent phosphorus, and 5 percent is potash.

These elements are always given in this order.

Nitrogen is the most important element in making plants grow. However, it is almost always scarce in soil because it is easily dissolved by water and washed away. The most common chemical compound of nitrogen fertilizer is *ammonium sulfate*. Organic sources for this include fish emulsions, cottonseed meal, dried blood, and sewage. Plants grown in partial shade require less nitrogen than sunlovers.

Phosphorus is necessary in plant cell formation and essential to the fertility of plants. Older leaves spotted with yellow, or with veins tinted purple, show a deficiency of phosphorus. Two good organic sources for phosphorus are fish meal and bone meal.

Also essential to plant growth, potash is needed by plants when they are maturing. It adds to the quality and endurance of plants, strengthening the tissues. Older plant leaves show a deficiency in potash when their margins turn purple. If unchecked, the leaves become brittle and drop off. Most potash fertilizers, which contain potassiums, come from minerals. Since they tend to run throughout the soil, only small applications are required every few months.

Organic fish emulsion fertilizers have always produced good results with container plants, but there is a marked preference for liquid concentrate chemical fertilizers. The general rule is to feed the plants every other week, applying according to package directions. Strong applications cause excessive leaf growth, fewer blossoms, and possible injury to roots. When using dry fertilizers, water the soil thoroughly to spread the nutrients into the root system. Fertilize less during winter when many plants are dormant or semidormant. Wait until plants show signs of growth before resuming regular applications.

Water the soil well before and after applying any fertilizer. Newly repotted plants do not need immediate feeding because they normally retain enough nutrients to sustain themselves. Never feed a sick plant.

Mulching

Nothing is more rewarding than summer mulching around the roots of outdoor plants. Mulch helps control weeds and retains moisture in the soil. Almost all plants benefit from the coolness a mulch provides. This is especially true of acid-loving plants with feeder roots near the surface. A mulched layer, thoroughly watered, keeps their roots from drying out during long, dry periods.

A few organic mulches that deepen and enrich the topsoil consist of leaf mold, peat moss, coarse sawdust, pine needles, wood chips, rice hulls, rotted manure, and straw. Nurseries carry many different brands of professionally prepared mulches. Organic mulches should be applied 1 to 2 inches deep for finely shredded substances, and as much as 4 inches in depth for rougher materials.

Inorganic mulches, such as crushed rock, gravel, or pebbles, provide weed control, good drainage, beautification, and permanency. Lay them *on* 2 to 3 inches thick.

Before mulching, remove leaves and other debris beneath a plant. This litter harbors pests. Sanitation is a factor in curbing the spread of some diseases common to plants.

Pruning

Some plants need a substantial amount of pruning, while others may require little or none. Reasons for pruning are as follows: to cut out dead, diseased, or injured wood; to prevent unshapely growth of a tree or shrub; to increase the quality and output of flowers or fruits.

Pruning is especially necessary in the case of deciduous shrubs and trees. It is best to prune off suckers as soon as they form. Cuts should be made directly against a trunk or branch, using sharp tools. Cover cuts over an inch in diameter with wax or a special sealing compound.

Flowering shrubs and trees must be pruned according to whether blooms appear in new growth or that of the previous season. For plants that flower in summer on new growth, pruning can wait until winter. Plants that bloom in the growth of the previous year should be pruned after flowering, thus enabling new shoots for flowering the next year. When pruning, make the cuts at a 45 degree angle, one-fourth inch from the bud. Use pruning sheers for small cuts, handsaws for large ones.

Pruning is often done to promote bushy growth, to keep a plant within an allotted space or at a desired height. Prune or pinch out the terminal buds on the branches to make the plant behave as you wish. This excess material can be used to propagate new plants.

Root pruning can increase the life of an overgrown plant and sometimes shock a poor blooming shrub or tree into full production. Use a large, sharp knife or a spade to cut a circle around the plant and cut the feeder roots. The diameter of the circle depends upon the size of the plant. For example, the diameter should not be less than 14 inches for a 4-foot plant. Cutting the feeder roots forces new roots to develop inside the root ball.

The best season for pruning is generally near the end of winter, before spring growth begins. Evergreen species subject to frost damage should never be pruned early.

INSECT PESTS

Insects can damage any part of a plant. With the exception of scale types, they move from one plant to another, overrunning them with infestation. Examine your plants regularly to avoid an all-out attack. If one does occur, it is best to consult your county or state agricultural extension service agent.

Chewing insects defoliate leaves. An example of these chewers is the prevalent caterpillar. They are the larvae of moths and butterflies, and can be found in many shapes and sizes, each specializing in their own dietary items. However, they all share alkaline stomachs. When they feed on a natural happening strain of bacterium called *Bacillus thuringiensis*, they quickly develop fatal ulcers. "Biocide" is a bacterial strain available commercially. To get rid of resident caterpillars, spray onto plants as directed. This control offers no hazzard to humans, nor does it kill beneficial insects. Praying mantises are helpful predators, for they feed upon caterpillars and other soft-bodied plant damagers.

Cutworms attack nearly all types of young plants, cutting off their stems at the soil line. They are usually found in the soil at the base of plants. Remove the worms from their source of food by turning the soil over lightly. A preventative is to place stiff cardboard collars around the stems of plants set out in the spring.

Earwigs are voracious night feeders that eat leaves, flower petals, and even climb into trees to feast on ripening fruit. These insects with biting mouth parts and sharp pincers at the tip of the abdomen hide under bark, debris or stones during the day. Trap earwigs by laying rolled-up newspaper on the ground. Shake them into a bucket of hot water or kerosene.

The Japanese beetle in the adult stage is harmful to roses and hundreds of other plants, chewing on both foliage and flowers. Lure them away by planting odorless marigolds nearby. Pick off beetle clusters and plunge them into a bath of kerosene.

Sucking insects, which dine on sap, are among the most harmful invaders of gardens. Since pesticides are highly poisonous to both human and beneficial insect life, it is much safer to control pests by organic methods and home remedies, some of which are herein discussed.

Aphids

Aphids may appear in many colors—green, brown, black, red, pink, or yellow—and are about one-eighth inch long. They will suck juices from all parts of a plant, weakening its growth. The honeydew which aphids secrete attracts ants. To spot them, look for black mold that grows on the honeydew, curled leaves, distorted buds, and stunted plants. Pick off and crush small infestations. Large colonies sucking beneath leaves and on stems can be washed away with a garden hose. A few ladybugs or praying mantises can devour a colony fast. Aphids attacking roots can be disposed of by pouring a pouch of tobacco on top of the soil and watering thoroughly.

Mealy Bugs

Mealy bugs are soft and fuzzy with cotton-white, waxy coats. They develop colonies at leaf and stem joints. Adults are about one-

fourth inch long. Their young are often seen in the crawler stage in the spring. Both distort leaves and stunt plant growth as they suck out sap. Destroy them outdoors by attaching praying mantis egg cases to twigs, or use the larvae of the lacewing. Control small colonies by touching each mealy bug with a cotton swab dipped in alcohol. Sponge larger colonies with mineral or vegetable oils. If possible, segregate an infested plant while treating it.

Mites

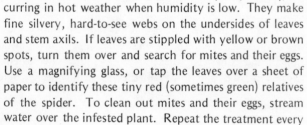

Red spider mites are one of the worst summer pests, usually occurring in hot weather when humidity is low. They make fine silvery, hard-to-see webs on the undersides of leaves and stem axils. If leaves are stippled with yellow or brown spots, turn them over and search for mites and their eggs. Use a magnifying glass, or tap the leaves over a sheet of paper to identify these tiny red (sometimes green) relatives of the spider. To clean out mites and their eggs, stream water over the infested plant. Repeat the treatment every few days until you are sure the pests are eradicated. Or, import lacewing larvae into your garden.

Cyclamen mites attack *Cyclamens*, geraniums and other plants. These microscopic pests make leaves turn dark, curled, or stunted. Try to control them by washing the foliage in soapy water several times. Then rinse in clear water a few more times. Destroy heavily infested plants.

Scale Insects

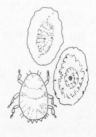

Various types of scale insects suck the juices of plants, remaining immobile beneath the scales they form. They attach themselves to bark, young branches, stems and leaf undersides. Their minute size prevents visibility before an infiltration becomes severe. Control outdoor attackers by releasing ladybugs at a plant's base in late afternoon. Clean stems and leaves of small plants in warm, soapy water. Then scrape off the shells with a thumbnail or small, stiff brush.

Thrips

Thrips are almost invisible, fast-moving, dark brown or black insects that distort foliage and damage plant tissue by sucking. They also ruin blossoms from the inside, often causing the partial opening of a bud. Leaves become curled, streaked, and silver-spotted where their juices are sucked. To control thrips outdoors, introduce lacewing larvae. Spray or wash small infested plants with a warm, soapy solution. Follow this with a rinse of clear tepid water.

Whiteflies

Whiteflies are tiny, mothlike insects that suck sap from the undersides of leaves and flutter off when disturbed. They are particularly attracted to soft-leafed plants. Whiteflies leave sooty deposits on leaves and stems. Their stationary, scalelike appearing larvae do most of the damage, feeding on sap until plant leaves turn yellow and drop. Check them outdoors by releasing ladybugs. For indoor plants, cleanse the undersides of leaves with a strong solution of warm soapy water and rinse with pure water. Try destroying the eggs and young by gently rubbing the backsides of leaves.

PLANT DISEASES

Most plants are practically disease-free when given the proper care. Just in case your plants should become victims of certain fungal or viral infections, a list of common diseases follows with recommended biological controls.

Powdery Mildew

Powdery mildew is a fungus disease that erupts to attack plants when the weather is hot and dry. Unsightly symptoms are gray-white powdery patches and mealy coatings on leaves, stems, and flower

buds. Roses and crape myrtle (*Lagerstroemia indica*) are among plants subject to this disease. Though plants seldom die, badly infected buds will likely fail to open. Dust with sulphur to control mildew. Prevent recurrence by moving the plants to a cooler spot. Good air circulation is important. Do not crowd them together.

Rust

Rust is a fungus disease which produces small, round pustules on leaves. These rupture and release brown, red, or yellow spores in the spring. Certain species and varieties of barberries, asparagus, and roses are susceptible to rust. The disease is fostered by plants being located in wet shady places, by plants being planted too close together, and by watered leaves not having the opportunity to dry out before nightfall. Correct the first two faults by relocating. Remove and burn diseased stems, leaves and flowers. Always look for rust-resistant plant varieties.

Leaf Spot

Leaf spot is a common disease which causes small or large brownish spots to appear on leaves. Infected leaves drop prematurely, but the spores produced are transferred to surrounding leaves. This fungus disease can be controlled by removing fallen leaves and burning them. Heavily diseased plants should be destroyed. Dewberries, blackberries and raspberries are only a few of the plants susceptible to leaf spot.

Wilt

Wilt is caused by a soil-inhabiting fungus that enters the plant through its roots. Ornamental shrubs and vines and many other plants are attacked by this disease. Leaves of affected plants turn yellow, red, or brown and fall prematurely. Stems and roots become streaked with reddish-brown lines. Older wood shows a brownish-green discoloration. Remove and destroy affected plants. Replace with wilt-resistant varieties.

Virus Diseases

Virus diseases of a number of kinds are apt to strike any plant, and are highly contagious. Common symptoms are spotted, circled or color-streaked leaves, stunted growth and yellowing foliage, followed by the death of the plant. Since no cures have been found for these diseases, it is necessary to remove the diseased plant at once and burn it. Aphids, leafhoppers, and other sucking insects are known virus spreaders. Watch for these and keep them under control.

HERB TEAS

In this book, the herbal teas mentioned are to be taken either hot or iced. Hot teas are made by pouring boiling water over the herb leaves or flowers and steeping them for 5 to 10 minutes to release the herb's essential oils. The general rule is to use 1 teaspoon of dried herb or 3 teaspoons of fresh crushed herb per cup of boiling water. For stronger flavor, it is usually better to use more of the herb than to steep the tea longer, because too much steeping makes a tea taste bitter.

Iced teas should be brewed stronger than hot teas, as melting ice dilutes flavor. Use boiling water and cool in the refrigerator before serving with ice. Herb teas may be sweetened with honey or sugar, and many are good mixed with other herbs. Be sure to try lemon slices, but never add milk or cream to an herb tea.

Solar energy herb tea is brewed by wrapping and tying the dried leaves in cheesecloth (gauze), then dropping the tied leaves into a quart-sized glass jar filled with cold water. Twist a lid on the jar and set it in the sun for 3 or 4 hours.

Root and seed teas should be simmered in water for 10 to 20 minutes to bring out their full flavor. Some have been recommended for generations as treatments for various ailments.

PART II

INDIVIDUAL PLANT SELECTIONS

SPECIAL PLANTS AND HOW TO GROW THEM

Described in this section of the book are many popular as well as lesser-known plants. All furnish quantities of material for propagation and are useful, with the majority being easy to propagate and grow.

Since ancestry plays an important role in plant requirements, the origin is pointed out in each description. This should aid the reader in selecting and growing suitable plants. Knowing where a plant came from is a clue to the kind of conditions and care that plant will need in order to perform its best.

Space does not permit descriptions of all the species, their hybrids and named varieties of individual plants. Therefore, use the enlarged index (explained in the Introduction) for information about other plants of the same species or genera. For complete details on propagation, plant care, fertilizing, watering, pest and disease control, readers will want to refer back to Part I of this book.

Following each of these special plant's cultural requirements, one or more methods of vegetative reproduction is suggested. This does not necessarily mean that other techniques are impossible. Everyone can enjoy the experience of experimenting with plant material, and thereby determine the method that works best for them.

The common names of plants are used in the text as much as possible, but for positive identification we must also include the tongue-twisting scientific names. These Latin or Greek names consist of two parts. The first is generic—the name of the genus, which is a plant group, or family, made up of different species—and the second word is the specific name of a given plant. For example, caraway is followed by *Carum carvi*. Knowledge of scientific terms is the key to a more accurate description of a certain plant, whereas a common name is often confusing because it may be only of local origin.

ACACIA, *Acacia spp.*

Origin: primarily Australia.

Native to dry regions, these ornamental shrubs and trees arrived in California not long after the discovery of gold. Presently, there are about twenty species that function well in places with mild winters. Acacias will not tolerate temperatures below 20° F. and therefore need to be kept in greenhouses wherever severe freezing occurs. Their quick growth, free-flowering habit, and resistance to drought compensate for their short life span of twenty to thirty years. Cultivated kinds vary from evergreen to deciduous—with feathery, divided leaves or a blade resembling a leaf—to low shrubs or tall trees, spiny or spineless. The small, fluffy, yellow or white flowers, usually in dense clusters, bloom profusely. Some perfume the air with fragrance. Kangaroo-thorn (*A. armata*), a spiny shrub 6 to 8 feet high, is widely grown as a decorative pot plant. Single, yellow, ball-like flowers bloom on its branches from February to March. Bailey acacia (*A. baileyana*), among the hardiest, is most often planted in the ground. It is a spineless shrub or small tree with lovely feathery blue-gray leaves. Branched sprays of fragrant yellow flowers appear abundantly in December, January and February.

Culture: Acacias prefer well-drained soil and full sunlight. Natural conservationists, they require only occasional, deep watering. Prune

after flowering to promote thick crowns. Several species of scale insects are sometimes troublesome, and two forms of caterpillars either roll the leaves or feed on these plants. Fungus diseases which may occur are leaf spotting, powdery mildew, root and tree rot. **Propagate** by hardwood, semi-hardwood or softwood cuttings. Transplanting must be done carefully to prevent injuring the long taproot that develops.

ALOE, *Aloe spp.*

Origin: northern and southern Africa.

These juicy succulents are often grown in pots or tubs as ornamentals for their showy leaves and clusters of branched or unbranched tubular orange, red, or yellow flowers. Aloes normally bloom during our winter, although some species produce blossoms throughout the year. Ranging in size from miniatures to trees, the plants form clumps of fleshy, pointed, and usually spiky-toothed leaves from which the flowering stalks arise. Of greatest interest is burn plant (*A. vera*), cultivated since Biblical times for the gelatinous juice inside the leaves. A piece of leaf is broken off and the juice rubbed on the skin to ease pain and heal minor cuts, burns, insect bites and other skin irritations. Eventually the grayish-green leaves become 1-2 feet in length. Flowers are yellow and bloom on a single stalk. For the grower who likes them on the large size, tree aloe (*A. arborescens*) is recommended, growing to 18 feet tall, and carrying huge clumps of gray-green spiny-edged leaves. Bright vermilion to clear yellow flowers bloom in long, spiky clusters from December to February. Perhaps most striking is *A. variegata*, partridge-breast aloe. Its clumped, dark green leaves are marked with bands of white, and clusters of dullish red flowers appear off and on all year.

Culture: Aloes will grow in fairly frost-free areas in almost any kind of well-drained soil. Where winters are harsh, they must be potted and brought indoors for protection. They need shaded conditions in summer; maximum sun in winter. Keep them evenly moist during active growth and treat monthly with a complete fertilizer. Watch for mealy bugs. Overwatering causes root rot, crown and stem rot. **Propagate** by severing offsets that develop around the parent plant, or by stem cuttings in those producing branching stems. Root cuttings in sand.

AMERICAN CRANBERRYBUSH, *Viburnum trilobum*
(V. americanum)

Origin: North America.

Also called highbush cranberry, this 8- to 12-foot deciduous shrub, found in woods, can be used as a substitute for the domestic cranberry that grows in bogs. A western species (*V. pauciflorum*) is similar. Altogether there are some twenty species in the United States. American Indians used an infusion of the gray bark for treating mumps and diuretic problems. The bark is still employed in medicine for its sedative properties. A mass of white, clustered flowers appear in summer, followed by large, reddish, 1-seeded fruits which have a sweetish-sour taste and a rather musty odor that diminishes as they age. The juicy berries are at their best cooked in pies, served as a side dish with meat, in making salmon-red jelly, conserves, and a wholesome beverage. A favorite fruit of birds, the handsome shrub should be covered with netting if the fruits are wanted for human consumption. The maplelike leaves turn brilliant red in fall, providing color to the garden. Plant this cranberry in a large container as a specimen, or in the ground for underplanting, screening, and as a hedge.

Culture: The American cranberrybush adapts to a wide range of climates. It grows well in heavy rich soils with ample water. It will grow in sun or shade, but performs best with some protection where summers are long and hot. Prune to prevent legginess. The plant is susceptible to aphids, spider mites, thrips and scale. To control leaf spot, pick off and burn all spotted stems and leaves. In late summer, bushes in shaded places may become covered with powdery mildew. **Propagate** by taking softwood cuttings in summer; by division; by layering.

APPLE, CRABAPPLE, *Malus spp.*

Origin: Eurasia and North America.

Since ancient times, the apple has been cultivated for its tasty fruits. Early American settlers brought with them seeds and cuttings of the better varieties of *M. pumila* from Europe. By extensive breeding, thousands of varieties have been developed which will grow almost anywhere in the United States. However, not all apples fruit well in all regions. Consult your county agricultural agent for varieties that do best in your area. Apple trees are appreciated for their scented blossoms in spring, for summer shade, and for the fruits they provide. Apples and crabapples contain vitamins, minerals and sugar. High in pectin, they are excellent to add to low pectin fruits to make jelly and jam. Varieties vary with respect to good eating, cooking, keeping, or food value. Besides standards that average 40 feet high, there are semidwarfs and dwarfs. Select a dwarf for container planting. They bear full-sized fruit and usually grow only 6 to 8 feet tall. To save space, train them, by espaliering, to decorate a fence, trellis or wall. If you want only one tree, be sure it can pollinate itself. Some of the most beautiful ornamental shrubs and trees are found among our native crabapples. The small, tart fruits are especially useful in jelly making.

Culture: All kinds of *Malus* prefer deep, well-drained, loamy soil enriched with organic matter. They need a sunny site and plenty of water. If leaves turn yellowish, scatter 10-10-10 fertilizer beneath the branches in early spring. About two dozen insects are known to attack the trees and fruit, the worst being the apple worm. Destroy the pests by spraying in late winter. **Propagate** by taking cuttings from grafted tops or softwood cuttings and rooting under high humidity conditions; by root cuttings taken from plants not over two or three years old; from suckers produced from the roots of standard-sized trees.

APRICOT, *Prunus Armeniaca*

Origin: China.

The decidious apricot tree has been under cultivation for thousands of years due to its delectable tangy-flavored fruit and its normally long life. After spreading to southern Europe, it was introduced to southern California in the eighteenth century by the mission fathers. Today there are an impressive number of varieties suitable for various regions in this country. Single white to pink blossoms blanket the 20- to 30-foot tree in early spring, but are frequently ruined for fruiting by late frosts. The golden-orange fruit is a good source of vitamin A, and develops its best flavor when fully ripened on the tree. It is often served raw or cooked as a dessert, or preserved by canning, freezing or drying. The dried fruit is a source of carbohydrates. Dwarf trees, growing to about 8 feet tall, will thrive in containers. They bear standard-sized fruit and many are self-pollinating. Check with your county agricultural agent for named varieties adaptable to your area.

Culture: Apricots succeed best in well-drained, light, loamy soil. Plant

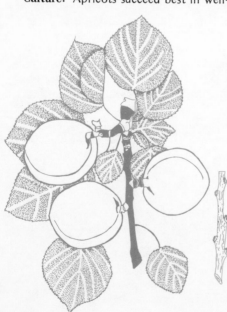

in a sunny but sheltered location to prevent frost damage to buds. In hot, dry weather water heavily once a month. Remove deadwood and overcrowded branches by pruning. Thin out excess fruit in midspring for larger sizes. In early spring, feed with 10-10-10 fertilizer along the tree's drip line. Spray for insect pests during the dormant season, before and after flowering. Tree and fruits are subject to blight and brown rot. **Propagate** either by softwood cuttings kept under a high level of humidity, or by hardwood cuttings of apricot or peach stock, followed by budding or grafting.

ARROWHEAD, Tule Potato, *Sagittaria latifolia*

Origin: North America.

Long before the coming of Europeans, Indian women from coast to coast waded into shallow waters of lakes and sluggish streams to dig out the tubers of the perennial arrowhead with their toes. Borne at the ends of long slender rootstocks, the starchy tubers furnish a nutritious and staple food, raw or cooked, the entire year. The plant is easily recognized by its sagittate (arrowhead-shaped) leaves that float above water or grow erect in marshy ground. Fragile white flowers appear in late summer extending 1 to 2 feet above the clustered leaves. Later, rounded heads of flat-winged seeds are produced. Although the whole plant is edible, the hard, overlapping, scaled tubers supply the most important food source. Small, rather bitter-tasting raw, they are good cooked like potatoes. They assume sweetness when boiled from 15 to 30 minutes, peeled, and eaten with butter, salt and pepper. Make them into a salad. Also try them baked in the oven for 30 to 40 minutes. If French fried or scalloped, they should be peeled sparingly to retain the greatest flavor and nourishment.

Culture: Arrowhead will grow beyond the edge of a pool if the soil is never permitted to dry out. It does well in an aquarium or tub large enough so the tubers can sink into a foot depth of mud. Mix 4 parts of garden loam with 1 part of well-rotted cow manure. Fill the container with the mixture to within 2 inches of the rim. Place the plants in the sun. Feed once a year by putting blood meal in a cloth bag and burying it in the soil near the roots. Aphids and leaf spot may be problems. **Propagate** by dividing the tubers in March or April and setting them upright in well-soaked soil; or from stem cuttings inserted in an open propagation box. Change the soil every third year. Harvest tubers in the fall.

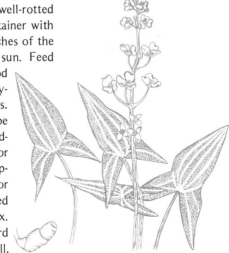

ARTICHOKE, *Cynara scolymus*

Origin: southern Mediterranean and northern Africa.

In ancient Rome the artichoke's bitter foliage was used in salads, but only in modern times have the flower buds been relished as food. Over the years the taste of this silvery-leafed perennial vegetable has greatly improved. Globe artichoke plants grow 4 to 5 feet tall and form large, deeply-cut leaves with a possible spread of 6 feet. On a patio it looks first-rate in a tub. Three plants will yield enough buds for a small family after the second year. Unharvested buds open into 6-inch purple-blue thistle blossoms, suitable fresh or dried for floral arrangements. Edible buds must be detached before they open into flowers; cut off approximately 1½ inches below the base of the bud. Artichokes grow best where summers are cool and winters frost-free. In colder winter climates, roots must be protected. This is achieved by covering them with a heavy mulch.

Culture: Artichokes need a well-drained soil rich in humus. Give them full sun, except in hot areas where they prefer afternoon shade. Feed them a balanced garden fertilizer in spring, and supply a top dressing of manure or compost with plenty of water until fall, when leaves yellow. Then cut the stalks to the ground. Control aphids by spraying the leaves

with a stream of water. If earwigs, slugs or worms lodge between the leaf buds, remove them by soaking the artichokes in warm salt water for ten minutes. **Propagate** by cutting off the suckers that pop up at the plant base during late fall or winter. Take cuttings with a piece of root attached and plant them in individual containers filled with rich soil. In early spring plant suckers 6 to 8 inches deep, setting them just above the ground and at least 6 feet apart. They can also be planted in individual tub containers.

ASPARAGUS, *Asparagus officinalis*

Origin: North Temperate Zone.

A hardy perennial vegetable, asparagus has been cultivated for hundreds of years for its delicious edible spears. It has become naturalized on the east and west coasts of the United States, particularly around salt marshes. Once established in a garden for two or three years, asparagus provides spears every spring for a period up to twenty years. Because of a massive root system, it is not recommended for containers. To save growing time, year-old roots are widely used. Eating the cooked spears, aids with the excretion of fluids, and is a good source of protein. They have also been used for gout and some rheumatic problems. The fernlike foliage forms a decorative background for other plants. In late spring the plant bears small, bell-shaped, greenish-white flowers. The fruit is an attractive red berry containing black seeds. Roasted seeds can be used to replace coffee beans, while crushed seeds can be made into a tea to relieve nausea.

Culture: Asparagus requires full sunlight and a deeply dug, loamy soil with a high humus content. For each root (or crown), dig a hole 8 to 10 inches deep and 1 foot wide. Fill with 2 inches of rotted manure and soak with water. Spread out roots and set so the tops are 6 to 8 inches below the surface. Cover with 2 inches of soil, add water to the crown, and apply 5-10-5 fertilizer. Gradual-

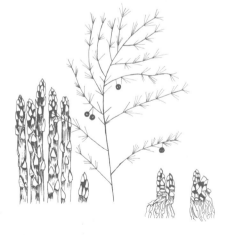

ly add more soil as the plant sprouts leaves. By late summer the hole should be filled to the original soil surface. Do not harvest the first crop. After removing the brown foliage in the fall, cover with a mulch of manure. Cut, at ground level, the spears that appear in the spring. Destroy snails, slugs and earwigs hiding in the mulch. Repel the asparagus beetle by planting tomatoes nearby. **Propagate** by division of mature dormant roots or crowns.

AUSTRALIAN LAUREL, *Pittosporum tobira*

Origin: China and Japan.

Although this species is not native to Australia, it does come from warm, dry regions. It grows beautifully outdoors in California and Florida, and makes a dense evergreen hedge. In temperatures that drop below zero, the leathery-leafed shrub or small tree must be container-grown and wintered indoors. It has a twisted trunk covered with smooth bark, and grows to an average height of 10 feet. The small, creamy-white clustered flowers, which appear in the spring, have an orange-blossom fragrance. In the fall, the brownish seeds split to show sticky orange-colored berries which provide food for birds. A smaller variety, *P. t. variegata*, has gray-green foliage edged with white. Both types respond well to trimming when young. They are attractive for screens, trained as free-standing trees, or as tubbed ornamentals for a patio or terrace. Very young plants can be trimmed to bonsai specimens, the glossy foliage adding a picturesque note to the crooked-stemmed trunk.

Culture: Australian laurel will grow in a variety of soils in the ground, but needs an acid mix when potted. It prefers sunshine, and flourescent lights help indoors. Though tolerant to summer drought, it shows

greener and richer growth when watered regularly and fed once each spring with a high nitrogen fertilizer. Mist once or twice a week if wintered indoors. Insect pests are aphids, mealy bugs and scale. Black sooty mold on leaves can be caused by mealy bugs. **Propagate** by taking stem cuttings 4 to 5 inches long in summer or early fall, inserting them in a moist, sandy compost. If potted, re-pot overcrowded plants every three or four years.

AZALEAS AND RHODODENDRONS, *Rhododendron spp.*

Origin: mountainous and hilly areas.

The general similarities of azaleas and rhododendrons pivoted the combining of these ornamental deciduous or evergreen shrubs into one genus. Mostly through the efforts of professional nurserymen, the once wild forms have been hybridized into vigorous, adaptable, foliaged plants, with magnificent flowers available in almost any conceivable color. Bursts of single or double blossoms can transform a commonplace yard into a brilliant focal point. The nectar-rich blooms lure bees and hummingbirds to the garden. Azaleas and rhododendrons are most commonly planted next to a house to accent a doorway, to hide a foundation wall, or to blend into a hedge with other shrubs. As a rule they grow slowly and make outstanding container specimens. Smaller species are satisfactory as ground or bank covers. The list of species and varieties is limitless. In your library, check the ratings for the hardiness of these delightful shrubs.

Culture: Since azaleas and rhododendrons are forest natives, provide a soil enriched with sand and coarse peat moss. Keep the soil moist and well-drained, and place the plants where they will be sheltered from wind and direct sunlight. Spray the foliage frequently for a cool, humid atmosphere. After blossoms fade, feed with an acidic plant food. Supply a mulch to keep the soil cool by day and warm at night. Principal pests are aphids, red spider and cyclamen mites, mealy bugs, scale, thrips and whiteflies. Possible diseases are leaf spot fungus and chlorosis (due to insufficient acid in the soil). **Propagate** by softwood cuttings of young growth in early spring, by division, or by layering.

BAMBOO, *Bambusa multiplex spp.*

Origin: Asia.

For impressive tropical effects, none of the tall grasses can excel the jointed-stemmed bamboos. They are valued aesthetically for their foliage, seldom flowering in cultivation. Most suitable for small gardens and potting are the clump bamboos which rise from underground rhizomes and make very little horizontal growth. Spreading slowly, they are easily kept within bounds. Requiring no more space than most shrubs, their widest use in this country is in landscaping. Native to tropical and subtropical regions, their permanent outdoor culture is restricted to the warmer parts of the United States. Elsewhere they must be kept in tubs placed indoors by bright light during the cold months. Since bamboos like potbinding, they can remain in the same tub for years. They should be thinned and clipped to show-off their dozen or more graceful stems. Chinese goddess and fernleaf bamboo are among the smallest species, reaching a height of 7 to 8 feet. They have ferny sprays of green leaves on solid stems. Each comes in dwarf sizes which can be controlled to 4 feet in height. Tender young shoots that form from the clumps break off readily and are excellent added to soups, stews and other dishes. In taste they resemble artichoke hearts, and their food value corresponds to that of an onion.

Culture: Bamboos are free from soil preference. For fast growth, water once a week and fertilize monthly. To restrain growth, keep on the dry side and never fertilize. Bamboos like bright light but little or no direct sunlight. Spider mites can be a problem, and dead canes left to decay may destroy an entire plant; cut them off at the rootline. **Propagate** by division of clumps in the spring, by softwood cuttings, or by layering.

BARBERRY, *Berberis spp.*

Origin: North Temperate Zone.

Patches of barberry can be found in woods and thickets across the United States. It is renowned for its brilliant fall foliage, golden-yellow clusters of flowers in spring, and for juicy, slightly acid, red berries used in making jams, jellies, preserves and a pleasant drink. Thorny-branched and low-growing, it makes a dense barrier hedge or an ideal single shrub. Indians and pioneers used the yellow root bark of California barberry (*B. pinnata*) and creeping barberry (*B. repens*) as a laxative, and to make a lotion for treating skin diseases. A bitter tonic made from the roots served as a blood purifier. From the leaves of *B. repens*, a tea was made to cure aches and pains. European barbary (*B. vulgaria*) is frequently found naturalized in the United States. Its root bark has the same qualities and uses. The clustered bright red berries should only be eaten when fully ripe. Fresh juice from the berries is sometimes used to toughen tooth gums. Numerous horticultural forms have evolved from these handsome natives, including dwarfs and minia-tures. Smaller varieties work well as borders for walks, mixed in flower beds, or placed in pots.

Culture: All species of barberry grow well in ordinary garden soil. A potting mix is best for smaller shrubs in containers. Most species can withstand heat, drought and fickle climates. Give them well-drained soil in sun or partial shade. Clip back for growth renewal. Deciduous species should be sprayed before spring growth begins to ward-off scale and other insect pests. If wilt appears through the roots, caus-ing premature defoliation, re-move and destroy the affected plant. **Propagate** either from cuttings taken in June—rooting them in moist sand, preferably in a shaded propagation box— or from suckers severed from mature plants.

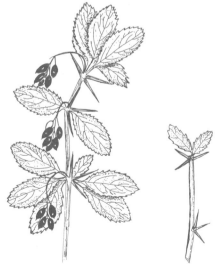

BEE BALM, Bergamot, Oswego Tea, *Monarda didyma*

Origin: North America.

Exuding a delightful mint fragrance, this handsome herb is native to moist areas of our eastern states. A similar species, *M. methaefolia*, can be found in pine forests and open valleys of the West. Various Indian tribes drank a tea made from the nettlelike leaves to reduce fevers, colds, headaches and sore throats. A stout, somewhat hairy perennial from 2 to 3 feet high, bee balm produces clusters of scarlet flowers surrounded by red-tinged bracts. The tubular flowers are a favorite of bees and hummingbirds. This plant is cultivated in herb gardens, and a fragrant tea can be made from its dried leaves, which is reputed to induce sleep. Fresh leaves rubbed onto exposed body parts is effective in deterring gnats and mosquitoes. The fresh leaves also lend a minty flavor to apple jelly and to fruit salad. This species includes a long list of named garden varieties, most of them easy to grow. They add color in several hues from June to September. All make fine cut flowers. Recommended are the scarlet Adam and the purple Prairie Night.

Culture: Bee balm enjoys any good, light, somewhat moist soil. It prefers partial shade, and needs trimming to keep it compact. The clumps of shallow roots spread rapidly and should be divided every three or four years. Problems may develop with mealy bugs, mites or whiteflies. Diseases are rare. Stimulate weak plants with compost, peat, or leaf mold. Spread on the ground surface and water thoroughly. **Propagate** by dividing root clumps in the spring. Fall divisions often kill clumps where winters are severe.

BLACKBERRY, Raspberry, Thimbleberry, *Rubus spp.*

Origin: North Temperate Zone.

Wild species of these delicious berries are abundant throughout most of the continent, finding a home in any climate where there is deep soil, ample water, and plenty of sun. Escaped domestic varieties are sometimes found along ditches. Erect or trailing, usually prickly plants, they flower and fruit during the second year of growth—after which the canes die. Each summer new canes are produced to bear fruit the following year. However, some species have perennial canes. The finely-haired leaves and roots are home remedies for diarrhea when made into a tea. Since Biblical times, leaves have been chewed for bleeding gums. The tender young peeled shoots are also edible. White flowers appear prevalently from June to September, followed by black, red or yellowish berries that are high in vitamin C. Fruits of wild blackberry and raspberry are eaten raw or used for pies, jellies, jams and freezing. When strained, the juice makes a refreshing drink. Thimbleberry, which grows at high elevations in damp woods and along streams, has maplelike leaves and is thornless. The berries are luscious fresh for desserts, or made into preserves.

Culture: Blackberries grow best when planted in deep, moist, well-drained soil in plenty of sun. Supplement the soil with a mulch when planting in the spring. Cut back newly planted erect types to within 6 inches of the ground. Mulch during summer months and remove stem tips to force the development of side branches. After harvesting fruit, cut out the canes that bore fruit. Let vining types creep until canes are 8 feet long, then cut off the tips. Tie the vines to posts and mulch heavily. After fruit bearing, cut back branches to 12 inches. Watch for caneborers, spider mites, whiteflies. Common diseases are rust, leaf spot and anthracnose. **Propagate** erect types by suckers and root cuttings, trailing types by layering and root cuttings.

BLUEBERRY, Huckleberry, *Vaccinium spp.*

Origin: eastern and western United States.

These deciduous and evergreen bushes are found in moist, high places, thriving under conditions befitting rhododendrons—to which they are related. Since blueberries are difficult to transplant, it is more practical to use the cultivated highbush blueberry (*V. corymbosum*) and its varieties. They function well almost anywhere with proper care. Blueberries offer glossy-leafed rounded bushes 5 to 6 feet tall, clusters of white bell-shaped flowers, and powdery-blue fruits that are rich in iron and great for pies, jams, jellies, preserves and syrup. In fall, the foliage is ablaze with vivid color. The long-lifed bushes are attractive grown as hedges or in containers. Plant two varieties to ensure fruit production. The closely related western evergreen huckleberry (*V. ovatum*) is similar, except that its berries are black. Sweet and flavorsome, they serve the same purposes as blueberries.

Culture: Both species need moist, well-drained, acidic soil and some shade in warm areas. After planting in early spring, they require a 4-inch-thick mulch to prevent roots from becoming exposed to the sun. Keep the mulch moist at all times. Apply a non-burning fertilizer such as cottonseed meal, using 4 ounces around young plants, 8 ounces around old ones.

Prune during the dormant season to remove weak growth and deadwood. The worst insect pest is a maggot that tunnels into the berries. To control it spray after the berries begin to mature. Mummy berry, a fungus disease which distorts the berries, is best controlled by heavy mulching. **Propagate** deciduous types by taking cuttings in November and placing them upright in boxes of damp peat over 70° F. bottom heat. Keep in an unheated shed until roots appear. Propagate evergreen types by using softwood cuttings. Layering is the easiest way to propagate both kinds.

BORAGE, *Borago laxiflora*

Origin: Corsica.

A pretty herb formerly used in medicine, perennial borage has become naturalized in various dry waste places of the United States. Growing about 2 feet tall, it is cultivated as a potherb, a rock garden plant, and as a drought-resistant groundcover. The small, bristly-haired leaves have a cucumberlike flavor and can be cooked as greens or used for pickling. Fresh, chopped leaves are good in green salads, omelets, or meat sauces. Dried or fresh leaves make an invigorating cold beverage or hot tea, high in organic calcium and potassium. A hot tea made from the fresh leaves has an ancient reputation as a lung tonic. To preserve the delicate flavor of the leaves, freeze them chopped or whole in sealed plastic bags. Lovely starlike opal-blue flowers nod downward in leafy clusters on the tips of this herb's branched stems which bloom throughout the summer. They make fine cut flowers or a colorful addition to a salad. Bees seek the nectar which they convert into an excellent honey. Borage repels worms that attack tomato plants if placed close to that vegetable. The related *B. officinalis*, an upright annual, serves the same purposes but must be discarded at the end of its growing season.

Culture: Borage accepts poor unfertilized soil, sun or shade, and a minimum of watering outdoors. If grown indoors, use a medium prepared for African violets. Cut back the leaves and stems severely to promote healthy growth before putting it outdoors in the spring. Repot in new soil before bringing inside in the fall. Set the pot in a sunny window and mist frequently. Outdoors, mulch the died-back plant with straw unless covered with a blanket of snow. Watch for mealy bugs, mites and whiteflies. Diseases are not a problem. **Propagate** by stem cuttings in the spring.

BOUGAINVILLEA, *Bougainvillea spectabilis*

Origin: Brazil.

In its tropical habitat this stout, woody, evergreen vine often reaches 100 feet, climbing rampantly to tree tops to make a flamboyant display of bloom throughout most of the year. It is one of the most widely planted ornamentals in the southern United States and in relatively frost-free areas of California. Outdoors the spiny, glossy-leafed creeper covers arbors, buildings, porches, fences and walls. Shrubby varieties grow successfully in pots indoors where the climate does not permit year-round outdoor culture. Three colorful bracts surround each flower, appearing in crimson, magenta, purple, and shades ranging from white, delicate pink, soft yellow to orange and brown. The variety Barbara Karst is hardy to 25° F. and is an improvement over the popular San Diego Red. Also it is agreeable to heavy trimming, which forces it into bushy growth.

Culture: Outdoor vining varieties will grow in ordinary garden soil, progressing rapidly to 50 feet unless pruned. Plant them in the sun and keep on the dry side during the blooming season to ensure masses of bright flaming color. In pots, use equal parts of sand, loam and peat; and keep the plant evenly moist. Give bougainvillea summer and winter

sun to encourage blooming, and mist daily in hot weather to promote humidity. Average house temperature will do. Feed monthly with a 5-10-5 fertilizer in early spring and summer. If indoor plants do not show good growth, apply an acid fertilizer. Mealy bugs, scale and leaf spot may be troublesome. **Propagate** by stem cuttings taken from April to June, cut into 3- to 6-inch lengths. Insert in a lidded propagation box to provide warmth while rooting. Cuttings may also be started under lights.

BURNET, *Sanguisorba minor (Poterium sanguisorba)*

Origin: North Temperate Zone.

Also known as salad burnet, this hardy perennial herb has been naturalized from coast to coast in the United States. Adaptable to most climates and found in bogs, open fields or sandy places, it grows from 12 to 18 inches in height. For centuries the astringent qualities of its deep-toothed leaflets were known and used to stop hemorrhages, and as an aid in healing wounds. The fernlike foliage was once added to warm wine and beer to make them more flavorsome. Today this bushy plant is cultivated by gourmet cooks for the young tender leaves which give a nutty, cucumberlike savor to salads, soups, sauces, cream cheese, or to serve as a garnish in place of parsley. Clean, fresh young leaves of burnet make a fine herb vinegar. Two-thirds of a glass jar is filled with washed, drained leaves. Then steaming hot cider vinegar is poured over the leaves, and the jar capped. After two weeks and several shakings, the salad dressing is ready to use. The delicately flavored fresh leaves also make a delightful tea. Tiny white, apetalous flowers crowd the short spikes at the top of the tall stalks but should be removed to promote development of new leaves and to prevent the plant from becoming woody. The plant itself should not be cut back more than half its size. Besides being a joy to any cook, planted in pots or in a mixed border, burnet serves as a year-round ornamental.

Culture: Burnet will tolerate any garden soil, but well-limed soil is best. It needs full sun, adequate water and good drainage to thrive. Possible insect pests are mealy bugs, mites and whiteflies. **Propagate** by division of roots in September, March or April. Since fresh leaves are a necessity, burnet must always be home-grown.

CALIFORNIA LAUREL, Oregon Myrtle, *Umbellularia californica*

Origin: California to Oregon.

This laurel can be easily identified by its ovalish, pointed-tipped evergreen leaves which have a strong spicy odor when crushed. In its natural environment, it either grows as a tall tree in forested canyons or as a huge shrub on windy hillsides near the coast. It rarely reaches more than 20 feet under cultivation. The dried leaves provide a perfect substitute for commercial bay leaves for use in soups, stews and meat dishes. California Indians made a tea from the leaves as a remedy for headaches and stomach pains. They also used the leaves in a steam bath to treat rheumatism; later the pungent oil from the leaves was rubbed thoroughly into the skin. Early white settlers mixed oil from the leaves with lard for rubbing the body to relieve aches and pains. In spring, yellowish-green flowers hang in dense clusters and give the plant an interesting yellow cast. They are followed by olivelike, thin-shelled nuts that are green and turn purple as they ripen. After roasting and cracking, the kernel becomes tasty and the original bitterness disappears. In gardens, this orderly plant makes an attractive shrub for hedges and screening. If thinned and kept small and shapely, it will adorn a container.

Culture: California laurel prefers moist, deep and rich soil, but is tolerant to less beneficial conditions. It favors dense shade, especially where summers are hot. Though tending to grow slowly, in time it becomes a shade-provider itself. Train as a shrub or tree by careful pruning. It is occasionally bothered by leaf-chewers and scale, but is disease-free. **Propagate** tardily by seeds. For quicker results try cuttings and layering.

CAMOMILE, *Anthemis nobilis*

Origin: Eurasia.

English gardeners in Shakespeare's day cultivated this perennial fernlike herb. Normally growing from 3 to 12 inches in height, the soft-textured bright-green leaves emit the pleasant fruity odor of apples. Camomile is commonly used in this country as a soil retainer for banks. It will grace a hanging basket on a patio or fill-in space around stepping-stones. With an evergreen and spreading habit, camomile makes a good lawn substitute that requires only occasional mowing. Equal parts of camomile and pennyroyal are said to soothe cuts and burns; to afford a chemical-free flea collar when sewed into a strip of cloth and secured around the neck of a cat or dog. The herb gardener grows the plant for the summer-blooming, clustered flower heads, which appear like small yellow buttons surrounded by white-petaled rays. These are much favored by bees. A rubbing oil or poultice made from the flowers is considered helpful for painful joints, swellings and calluses. Since colonial days, camomile tea has been a home remedy for indigestion, colic, fever, spasms, and for calming overactive children. Flower heads in full bloom are gathered and dried. Steep them from 3 to 5 minutes for best results. To add a spicy flavor to the tea, combine with freshly grated ginger.

Culture: Camomile grows readily in any well-drained garden soil. It prefers the sun but will endure a light shade. Apply water moderately. Few pests bother this pungent plant. About the only disease that attacks is mildew, caused by too much shade. **Propagate** by division of roots in spring or fall.

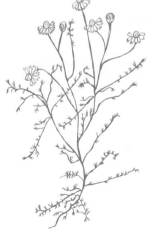

CARAWAY, *Carum carvi*

Origin: North Temperate Zone.

Seeds of caraway were found among the ruins of the Lake Dwellers of Switzerland, proving that the virtues of this aromatic herb have been known since ancient times. For centuries licorice-flavored caraway seed was used as a cure for upset stomach and indigestion. The biennial plant grows wild in most parts of the United States, thriving in open ground waste places. There is little difference between it and cultivated caraway. Finely divided dark green leaves extend 1 to 2 feet high the first year and the foliage remains throughout the winter. The following spring umbrellalike clusters of creamy white flowers rise above the foliage. By midsummer the flowers develop into brown seeds. Cut the stems before the seeds fall, tie in small bundles and hang in a warm but airy place. Let the seeds drop into a paper-lined tray. Store the dried seeds in airtight jars to use for flavoring pickles, bread, cakes, cookies and cottage cheese. One plant produces about 4 tablespoons of seed. The tiny green buds can be sprinkled on cooked cabbage, boiled beets or potatoes, to give them an aromatic taste. Fresh or dried seeds add a dramatic touch cooked in fruit sauces and baked apples. Ground dried seeds are good to flavor meat, poultry and stuffings. Young caraway leaves make an attractive garnish. The white

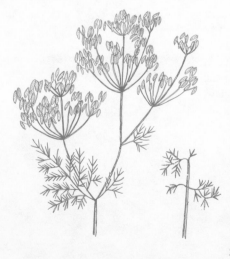

mature root looks like a carrot and contains the same food values. Eaten raw, boiled or baked they are flavorsome and nutritious.

Culture: Caraway grows best in moderately rich, well-drained soil. Plant in sunlight or partial shade and water regularly. Insect pests and diseases are negligible. Because the plant dies after the seeds ripen, **propagate** by stem cuttings taken during the spring off the second year's growth.

CATNIP, *Nepeta cataria*

Origin: Eurasia.

Ever since Greeks and Romans grew catnip to please their cats, this aromatic herb has been popular with cat lovers. Cats like to roll in the mint-scented leaves and to eat them dried. Few people know that this perennial repels rats. After the 2- to 3-foot-tall plant was brought here by colonists, it escaped cultivation and began growing wild in woods and thickets throughout the United States. A vigorous spreading woody herb, with soft green, heart-shaped leaves, catnip makes a fine groundcover if kept clipped. Untrimmed, it will form a splendid background for lower growing plants. Grown in pots, the plant can be forced to stay within bounds. Oil in the stems and leaves is high in vitamin C, never losing its strength when dried and stored in airtight jars. A delightful warm tea made from the dried leaves is considered effective for infant colic, upset stomach and acid indigestion. The tea can be used to calm nerves and induce sleep without side effects. Cold catnip tea served before meals tends to whet the appetite, while fresh leaves rubbed on meat before cooking deepens the flavor. A spray made from the dried leaves controls insects infesting houseplants. Fresh catnip leaves, crushed and sprinkled along ant trails, repels these pests. The clustered, lavender-to-white, summer-blooming flowers attract bees. Who can say catnip is solely for the pleasure of cats?

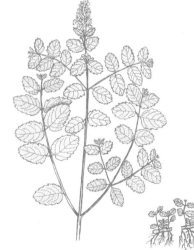

Culture: Catnip thrives in light soil with lime, and with sunlight or partial shade. Indoors it needs good light. Keep it moderately moist and feed monthly with a 20-20-20 fertilizer. Protect from heavy frost by bringing it indoors in the winter. Seldom is catnip troubled by insect pests and diseases. **Propagate** by division of roots in the spring or fall; by stem cuttings taken in the summer.

CENTURY PLANT, *Agave americana*

Origin: the Americas.

Numerous species of century plant grow wild on dry slopes and in deserts from the southwestern United States to the equator. It can be distinguished from aloe by the hooked spines along the leaf edge and the cruel spine at the tip. From the gray-green, lance-shaped leaves of the century plant, the Aztecs manufactured a paper resembling papyrus. They also wove a twine from the tough leaf fibers. At one time desert Indians used the buds from the long ascending stalks for food and drink. The clustered seeds from the flowers were beaten into flour. Today these plants are considered emergency food only. Contrary to general belief, century plants take 10 to 50 years to bloom. After blooming they die, leaving offsets at their base. A number of striking horticultural forms have been developed from the wild plants. Among the largest is *A. a. marginata*, with leaves edged with yellow. It makes an impressive ground or tub specimen where space is available. A medium-sized species, *A. attenuata*, with soft unarmed gray-green rosettes, is widely used in landscaping and in containers. The most elegant of the small, slow-growing agaves is *A. victoriae-reginae*. Its stiff narrow green leaves, marked with white edges, form a compact clump about a foot across when fully grown. Small, young agaves are excellent as potted plants for the patio, porch or in a sunny window.

Culture: Century plants will survive in poor ground soil, but in containers they do best in a mixture of 2 parts loam and 1 part sand. Protect them from the hot sun and from frost. Apply water when the soil surface is bone dry. A light dusting of bone meal is a good yearly fertilizer. Search for mealy bugs and scale. Excessive moisture causes rot to occur. **Propagate** by removing offsets at any time. Let the soil remain dry several days after potting.

CHECKERBERRY, Wintergreen, *Gaultheria spp.*

Origin: North America.

Long before the coming of the colonists, the American Indians knew the value of checkerberry (*G. procumbens*). In fall months they would collect these dark green, shining, oval leaves from damp, cool forests. They showed the colonists how to brew an herblike tea from the leaves of this small evergreen shrub, to treat headaches and various internal ailments. According to folk medicine, the liquid steamed from the oil in the leaves was used as a liniment for external body aches and pains. A woody plant, its foliage has the pleasant taste and smell we know of as wintergreen. The oil of the plant is closely related to aspirin. Dried and powdered leaves can be used as tooth powder. The waxy, white, bell-shaped flowers bloom in the summer, borne from the leaf axils. They are followed by tiny red berries. Wintergreen flavored, the berries are edible raw or cooked. For birds and deer, they are an important winter food. One western species, *G. ovatifolia*, has a marked similarity to the eastern plant. Both have creeping stems with upright branches, making them good subjects for groundcovers, trailers over rocks, hanging baskets, and companions for azaleas, rhododendrons and ferns.

Culture: This plant must have a humuslike soil enriched with sand and peat, and covered with a mulch of leaf mold, pine needles or peat. Plant in a cool, shaded place where there is filtered sunlight and plenty of moisture. In hot summer climates, spray the leaves daily with water. Cut out deadwood in April. Rarely is the plant attacked by insect pests. Principle diseases are leaf spot and powdery mildew. Pick off and burn infected leaves. **Propagate** by stem cuttings when the wood is half-ripe and flexible.

CHINQUAPIN, *Castanopsis spp.*

Origin: Pacific Coast.

The beautiful chinquapin, a close relative of the American chestnut, is a slow-growing evergreen shrub or bushy tree which once thrived in open woods at low elevations across the United States. Blight, a fungus disease, destroyed the eastern natives whose foliage was used by the Cherokees to make an external wash to treat chills, fever, headaches and sweating. The hardy western natives abound in northern California, Oregon and Washington. A few species are cultivated in home gardens. The attractive dark green tapered leaves are glossy above with golden feltlike hairs beneath. In summer, strong-smelling, creamy white flowers appear and are borne in fluffy catkins. In the fall come nuts enclosed in small prickly burrs. The clustered burrs open to drop one to three nuts to the ground. Kernels inside the nuts have a sweet, rich flavor and food value. They contain 45 percent starch and 2.5 percent protein. The tree form *C. chrysophylla*, called giant chinquapin, grows from 20 to 25 feet tall under cultivation, has furrowed bark and spreading branches. Golden chinquapin (*C. c. minor*) is shrubby and grows 3 to 15 feet in height. *C. sempervirens*, bush chinquapin, is another shrub type but reaches only 8 feet in height. Very sturdy, it stands up well under a blanket of snow. Plant this one in a container or use as a hedge.

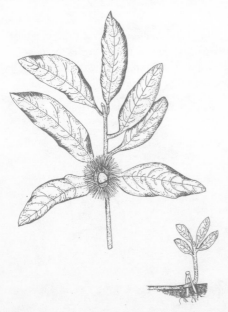

Culture: Chinquapins will flourish in full sun and poor soil. They require little attention beyond an occasional watering once established. A straw mulch is helpful under young plants. To restrain growth, prune after flowering. The plant is bothered by few insect pests and is now blight resistant. **Propagate** by stem cuttings, suckers or layering.

CHIVES, *Allium schoenograsum*

Origin: Eurasia.

A hardy perennial herb, this member of the onion family is found growing wild in rocky pastures and mountain areas of its native homelands. It was introduced to the New World by early settlers who cultivated the plant for both culinary and ornamental purposes. Chives is far smaller than the typical onion and the bulbs form in dense clusters. Its hollow, grasslike leaves, with an onion scent, can reach 2 feet in height, but much less if kept potted indoors. Leaves should be cut and used often to encourage new growth. Finely chopped leaves give a delicate, onionlike flavor to salads, mashed potatoes, omelets and soups, In addition, the leaves offer a delicious taste to chicken dishes, hamburger, spaghetti and hot vegetables. The ball-like, clustered purple flowers appear in the spring atop thin stems. They are favorites in flower cookery, flower arrangements, and as garnishes. Seeds furnish seasoning for cooked salads. Plant chives in *sets* to edge flower or vegetable borders, as the herb is helpful in repelling aphids and Japanese beetles. If potted, bring the plant indoors and place on a sunny window sill to prevent dormancy where winters are harsh. Or, freeze chopped leaves for winter use.

Culture: Chives grow in ordinary garden soil, but when potted they need rich, slightly acid, sandy soil. To make the leaves more tender, provide plants with afternoon shade. Some moisture is required. A built-in immunity guards the plant against injurious insects, but bulb rot may develop. The disease affects bulbs, stems and flower stalks. Burn infested bulbs to stop the disease from spreading. **Propagate** by dividing bulbs every 2 to 3 years in late spring or early fall, planting about 8 to 10 to a pot.

CITRUS, *Citrus spp.*

Origin: tropical or subtropical Asia, Indo-Malaya.

The natural homes of citrus shrubs or trees limit their normal production to the warmer regions of the United States. The most cold-tolerant types remain hardy to 20° F. Dwarf citrus, growing 4 to 10 feet in height, can be enjoyed by gardeners in colder areas if the trees are confined in tubs that can be taken outdoors in the summer and brought indoors during the winter. The citrus group includes lemons, limes, grapefruits, kumquats, oranges, tangerines, and their various hybrids, which in part, consist of limequat, tangelo, tanger and sour oranges. All have glossy, oval leaves and bloom delightfully fragrant white blossoms. Their colorful fruits, edible raw or cooked, are rich in vitamin C. Varied citrus species and their varieties have fairly specific climate requirements to produce the best fruit. Consult your county agricultural agent for information regarding the citrus suitable for your area.

Culture: Citruses require total sunlight, and a well-drained soil to prevent root rot. Plant in raised beds if the soil is especially heavy. For containers, use equal parts loam, soil and peat. Irrigate a newly planted specimen twice a week to establish a strong root system. After that, water thoroughly every other week during hot summer weather. Make a water basin around the plant and spread a mulch in it to keep the soil

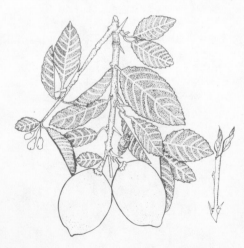

cool. To feed, apply a high nitrogen fertilizer, or a packaged fertilizer especially prepared for citrus, in late winter and in June and August. Harmful insects are aphids, mealy bugs, scale and spider mites. Yellow leaves may occur due to an iron deficiency, or overwatering. If the plant is not getting too much water, apply iron chelates to the soil. **Propagate** by rooting cuttings of half-ripened wood in spring or fall.

COLEUS, Painted Nettle, *Coleus blumei*

Origin: Java.

Brilliant color and velvety foliage characterize this shrubby ornamental from the tropics. Coleuses ordinarily stand erect, growing from 1 to 3 feet high. Among the countless hybrids and named varieties, some are creeping forms, desirable for hanging baskets. Variegated types have long been known and prized for their wide range of coloring. Multi-hued leaves come in every color imaginable, ranging from cream, chartreuse, pink, purple, red or brown. Edges may be ruffled, fringed or scalloped. Blue or lilac flowers, borne in terminal clusters, should be removed if the plant is to grow luxuriantly. For a gorgeous summer display, plant coleuses in beds or in a window box on the shady side of the house. This rather weak, non-edible herb is too tender to survive severe cold. To grow as year-round specimens, plant them in pots and bring indoors before frost threatens. Transplant whenever the roots fill the pot—until finally lodged in 10-inch pots. Repotting is sometimes necessary two or three times a year. Stake and tie the large branching plants to prevent their stems from snapping.

Culture: Coleuses need rich, loose, well-drained soil. Light should be bright, but too much sun produces less vivid coloring. Turn potted plants frequently to develop symmetry. Keep the soil evenly moist and feed mature plants twice a month with a high nitrogen fertilizer. Pinch stems regularly to encourage branching and compact growth. Destroy slugs and snails that attack outdoors. To keep healthy, control aphids, mealy bugs, spider mites and whiteflies. Leaf drop is caused by insufficient water. **Propagate** by placing stem cuttings in water or in a rooting medium any time of the year. Cuttings of firm, young shoots root readily in a potting mix.

CORIANDER, Chinese Parsley, *Coriandrum sativum*

Origin: Asia and southern Europe.

Coriander is a perennial herb that has been cultivated for thousands of years, particularly for the medical properties in its seeds. Although the delicate, fernlike leaves and stems are rather bitter tasting, they are frequently featured in Latin and oriental salads, soups and rice dishes. Young leaves and stems are also used to flavor meat, poultry, and as a garnish like parsley. During summer, flat clusters of pinkish-white flowers appear, followed by brownish seeds. The seeds have an unpleasant, musty odor until they ripen, after which they take on their spicy aroma. Planted along a flower or vegetable bed, the smell of the unripe seeds repels aphids. Both leaves and seeds become sweet when dried and develop the taste of lemon peel and sage. Containing vitamin A, whole dried seeds are a healthful ingredient to add to sauces and pickling liquids. A tea may be steeped from the seeds to relieve dysentery. Ground seeds introduce a spicy flavor to candy, cookies, breads, pastries, beans, stews and cooked vegetables. Oils of the seeds provide scent for soap. For drying, the ripe seeds must be harvested when fully brown, then stored in an airtight container. To get enough seeds for the average family, a 6- to 8-foot row of coriander is necessary.

Culture: Coriander grows best in rich, well-drained soil, requiring full sun and regular watering. It needs protection from strong winds and frost damage. The plant is not damaged by insects nor by diseases. **Propagate** by stem cuttings to obtain the quickest results.

CRAPE MYRTLE, *Lagerstroemia indica*

Origin: China.

Few shrubs or small trees display as much color during summer as crape myrtle. Until mid-September, each branch and twig is adorned with fluffy clusters of fringe-petaled flowers of snow white, rich lavender, deep red or watermellon pink. The oval leaves are bronze-colored when they unfold in spring, but before dropping in the fall they become yellow, orange or red. The smooth brown or gray bark slowly falls off, disclosing pinkish bark beneath. The bark patterns, twisted branches, and small brown seed pods are attractive during winter months. Tree forms reach about 30 feet in height if trained and pruned to grow in a single trunk. Smaller bush types, growing 5 to 7 feet tall, are practical for limited space. Plant them as informal hedges, or in containers to brighten up a porch or patio. In severe winter climates, store a pot-grown plant in a frostproof place. Should the tops become frostbitten, prune them in the spring to promote new growth.

Culture: Crape myrtles grow best in the sun, and where summers are hot. Well-drained soil supplemented with peat and leaf mold is advisable. Apply water deeply but infrequently. Feed two or three times during the growing season with a 5-10-5 fertilizer. Prune out weak shoots and deadwood before spring growth starts, or to control the size of tree forms. Aphids are perhaps the worst insect pests. Powdery mildew indicates either that a plant needs more space for air circulation, or that the location is too moist and shady. **Propagate** by taking 12- to 15-inch softwood cuttings in late spring or summer; by semihardwood cuttings in late summer; or from hardwood cuttings of dormant growth in late fall or winter. Hardwood cuttings root readily if inserted 6 to 8 inches deep in moist sand and kept in a protected spot until spring.

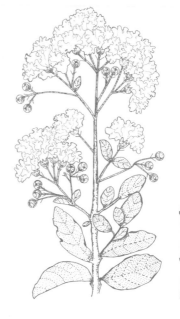

CURRANTS, GOOSEBERRIES, *Ribes spp.*

Origin: North Temperate Zone.

There are some eight species of wild currants and gooseberries grown across the United States and Canada. Normally they are found in open or shaded moist woods or hillsides, growing 3 to 5 feet high along streams or bogs. All produce fruits which may be eaten raw, cooked, or dried. The neat, nonspreading bushes offer maplelike evergreen foliage, fragrant spring flowers, and fruits that ripen in midsummer for making pies, tarts, puddings, jellies and jams. Varieties which can be eaten raw are an especially good source of vitamin C. Gooseberry bushes are thorny, whereas currants are not. Both send up a crop of new shoots from the ground each year. They bear fruit at the base of year-old wood, and on spurs of wood two and three years old. Because currants and gooseberries are alternate hosts to white pine blister rust, their planting is prohibited in many states. Your state agricultural extension office gives information about this.

Culture: Currants and gooseberries require a moist, heavy, well-drained soil mixed with organic matter. Mulch them to keep their shallow roots cool. If the mulch contains nitrogen, feed yearly with bone

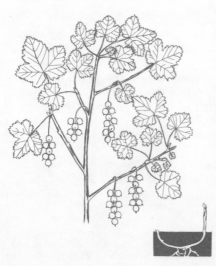

meal. Provide shade in hot-summer areas, and plant in the sun in coastal areas. Prune every few years in late fall to remove older canes and weak growth. Wilting leaves denote the presence of cane-borers, so cut back an infested cane until the borer is found and destroy it. Burn bushes that show orange-colored blisters under the leaves. Mildew may also occur. **Propagate** by layering, by hardwood cuttings, or by mounding. Gooseberries root easiest by mounding.

DATE PALM, *Phoenix spp.*

Origin: Asia, north Africa.

These elegant feathery palms were introduced from the Old World into the southwestern United States in the twentieth century, and are cultivated commercially for their delicious clusters of fruit. In most parts of this country the production of drooping, yellowish flowers and edible dates is limited to frost-free regions. Because of their attractiveness, date palms are widely used as houseplants, to be summered outdoors as tubbed specimens on porches and patios. For years they will thrive in the same container, requiring little attention. The *Phoenix* genus includes *P. dactylifera* with gray-green, waxy needlelike leaves. It adapts to a container when young and makes a charming indoor plant in winter. Pigmy date palm (*P. roebelinii*), with its delicate arching fronds, is a dwarf with a single trunk. It eventually grows to 6 feet, but may take fifteen years to reach that maximum height. Under favorable conditions the mature plant yields edible dates, which are about half an inch long.

Culture: *Phoenix* palms will grow in well-drained clay or sandy soil, but do best in food-fertile soil with one-half cup of bone meal added to a 10-inch pot. Light should be bright; however, do not place them directly in the sun, whether indoors or out. Keep the soil quite damp from April to October, but they require little water throughout the rest of the year. Spray the foliage daily in the hottest weather to remove dust and insects that hide in the leaf stems. Feed established plants with a 5-10-5 fertilizer monthly during the active growing season (summer), but not during dormancy. These palms are attacked by mealy bugs and scale. Crowns buried deep in the soil can cause root rot. **Propagate** by cutting suckers from mature plants in the spring. Pot them in porous soil in 5- to 6-inch containers.

DOGWOOD, *Cornus spp.*

Origin: North America.

Our beautiful American dogwoods are found in moist woods beneath trees or along streams throughout the eastern and western regions. These hardy shrubs or small trees are attractive year-round in landscaping. They usually display an abundance of eye-catching petallike bracts and brightly-colored glossy fruits. When snowdressed, the branches offer a breathtaking show. Indians boiled the astringent inner bark to arrest diarrhea, fevers and colds. From the boiled inner bark they made compresses to cure hemorrhoids. Dried bark has been used as a substitute for quinine. Bunchberry (*C. canadensis*) is a 6- to 9-inch-high creeper which grows wild almost nationwide and blooms white flowers from April through June. It has clusters of small, scarlet berries that are edible raw or cooked. Cornelian tree (*C. florida*) forms a tree that grows as high as 20 feet. Small white, pink or rose flower bracts nearly cover the gray branches before the leaves unfold. The shiny red fruit is an edible berry. Autumnal foliage is a brilliant red. Pacific dogwood (*C. nuttalii*) is a spectacular western tree reaching 50 feet in height. Shining white, sometimes pinkish, flower bracts develop into gleaming clusters of orange or red edible fruits, and the oval leaves turn to lovely reds and yellows before falling. Redtwig dogwood (*C. stolonifera*) is native to most of North America and is noted for its showy red winter twigs. It is a shrub about 15 feet high, with creamy white

flowers blooming in profusion among the leaves from summer until fall. The fruits are white or blueish.

Culture: Dogwoods need good drainage, rich humus, partial shade and frequent watering in dry weather. A heavy mulch is helpful. All remain hardy below zero. Trim branches of shrubby types where they touch the ground. Watch for insects and diseases that enter injured bark. **Propagate** by hardwood cuttings or by layering.

ELDER, ELDERBERRY, *Sambucus canadensis*

Origin: North America.

American or Canadian elder is usually a branching shrub from 4 to 12 feet high. Deciduous, it commonly grows along partially shaded woodland streams, ditches, roadsides, and at the edge of bogs throughout North America. Because of the wide Indian usage, elder became important in home remedies of white settlers. The yellowish-gray bark was boiled and prepared as a wash for imflammations. From the leaves and tiny white dried flowers, a wash was made to treat wounds. The dark purple berries were made into a relaxing, laxative drink. Elder is still cultivated for its clusters of juicy round berries which are high in vitamin A and are good in pies, cobblers and catsup. Juice from the tart berries is made into a beverage, and provides a delicious jelly combined with equal amounts of apple. The scented flowers are made into tea, or used for flavoring muffins, pancakes or waffle batter. They are also used to make an ointment for burns and a skin cream. The green buds can be pickled, while the leaves are believed to repel insects. Prevalent on the west coast is *S. caerulea*. This elderberry has especially large, edible blue berries. *S. racemosa*, another elderberry found in the West, has cut-leaf foliage and bright red berries.

Culture: Elder is hardy, growing well in cold climates as an effective screen or windbreak. Soil should be fertile and always damp. Shade it from direct sunlight. Prune the fast-growing plant each dormant season to keep it dense and shrubby. Pests are aphids and caterpillars. If currant cane borers appear in the hollow stems, destroy them. **Propagate** by taking 6- to 12-inch hardwood cuttings and rooting in sand; by division; or by separating new shoots from the parent plant.

EVENING PRIMROSE, *Oenothera biennis*

Origin: North America.

A great number of native evening primroses are distributed across the United States, but this is the biennial known by various common names. A rather coarse, erect, branched herb, it is found in dry pastures and open fields, growing 3 to 4 feet tall. It was one of the first edible plants to be sent to the Old World for cultivation. Boiled or French-fried, the fleshy, branching roots are sweet and nutritious during the first year of growth. Their size depends much upon climate. From the astringent roots, home remedies for coughs due to colds, and an ointment for treating minor skin problems were developed. The first year the broad rosettes of leaves with toothed or wavy margins are 1 to 6 inches long. Rosettes, along with the reddish stems and shoots, may be used as greens. Arising from the second year's growth are long-stemmed, large silky-petaled yellow flowers that open around sundown to emit a sweet-smelling fragrance. Their color and odor attract sphinx moths at night, upon whom they depend for fertilization. In the morning these lovely summer bloomers close their petals, which often wilt and fall off in a day or so. The plant dies after the hairy seed capsules are produced.

Culture: Evening primroses are easy to grow if given well-drained sandy or loamy soil in a sunny location. In a border or rock garden they are never aggressive. For rapid growth, apply liquid fertilizer occasionally. Drought-resistant, they require little water. Another advantage of these natives is that they are virtually pest and disease free. **Propagate** by taking stem cuttings from the first year's growth, or by root division.

FERN, BRACKEN, *Pteridium aquilinum*

Origin: Northern Hemisphere.

These coarse, feathery ferns with long-creeping rhizomes are widely scattered across the United States and are common in moist places at lower elevations, or as groundcovers in forests at higher levels. The usual height is 16 to 32 inches. Bracken is one of the hardiest fern types for cultivation, whether planted to dramatize the north corner of a house or tubbed on a shaded patio. The blackish rhizomes can be baked or boiled until soft and doughy, peeled, and the nourishing insides eaten, or else used to thicken and flavor soups. Only young ferns, less than 8 inches tall, can be safely eaten as a vegetable. Full-grown fronds become tough and poisonous. The young fronds, boiled in salted water until tender, have a delicious, asparaguslike taste, especially after salt, pepper, and butter are added. In early spring, the rhizomes send up coiled sprouts called fiddleheads, which are edible raw or cooked. Served raw in a tossed salad, they affix a pleasant sweetish taste. Fiddleheads may be stored by freezing or canning.

Culture: Bracken thrives in deep, moist, highly organic soil which is slightly acid. A pine needle mulch helps retain moisture. Full sunlight should be avoided to prevent fronds from burning. Frequently spray down the foliage to keep fronds clean. Cut off dead and decaying fronds and brittle leaf edges. Watch for aphids, mealy bugs, scale, slugs and snails. **Propagate** by division of rhizomes in the spring when warm weather approaches. Put the rhizome sections in small containers filled with a light potting mix, covering them lightly with the medium. Place in a warm and humid, but not sunny spot. Keep the developing plants moderately moist and the temperature as constant as possible.

FIG, *Ficus carica*

Origin: Mediterranean Region.

The deciduous fig tree has been cultivated for thousands of years for its bountiful, sweet fruit. Its fairly rapid growth produces a many-branched tree, not over 30 feet high. It can be trained low for a shrubby effect and grown in a large tub. Fig trees like long hot summers and survive winter temperatures to 15° F. without damage. In colder areas, the tree must be completely wrapped for protection. Many varieties bear fruit without pollination and are suited for certain regions. Two crops of figs can be expected from any established tree. One appears in early summer from the buds of the previous season's growth; the second in late summer on the current season's growth. Figs vary in color from greenish-yellow to dark purple, and have a mild laxative effect. All may be served raw as dessert or made into jam or preserves, but they are most commonly eaten dried. Roasted figs cut in half, make excellent poultices for boils. The acrid, milky juice of leaves and stems has been applied to successfully remove warts. However, some people are allergic to the milky juice.

Culture: Fig trees are not particular about soil, but good drainage is necessary. They prefer full sun and shelter from strong winter winds.

Spread a thick mulch of wood chips around newly planted trees to prevent the roots from drying out. Once established, they become quite drought-resistant. Fig trees seldom need fertilizer, nor are they much troubled by insects or diseases. When fruit is ripening, protect from birds with a netting cover. **Propagate** by taking 12-inch hardwood cuttings gathered in late fall from the previous season's growth. Insert them in moist soil so only one bud shows. Come spring, a new tree will begin to grow from each cutting.

FILBERT, Hazel, Hazelnut, *Corylus spp.*

Origin: North Temperate Zone.

These deciduous shrubs or small trees are grown for their flavorsome nut kernels and are useful as hedges and shade trees. *C. avellana* and *C. maxima*, along with their varieties, are filberts of European extraction. They produce the larger nuts, but tend to be less hardy than our native American hazels that grow near the edges of woods, damp slopes and streams, in the eastern and western United States. Beaked hazel (*C. cornuta*) has ovalish, dense, double-toothed leaves that turn yellow in the fall. Greenish-yellow male catkins appear which hang on the plant throughout winter. Tiny female flowers bloom in spring in separate clusters near the top of the branches. They develop into bristly seed husks with long beaks. After the nuts drop to the ground in the fall, gather them and plunge them into water, removing the floaters. The kernels inside the shells are crisp when eaten raw, and may be used for flavoring cookies, breads and candies, after they are chopped.

Culture: Filberts and hazels need deep, fertile, moist, well-drained soil. Roots extend down to 12 feet at maturity. For the first two or three years, protect them from too much sunlight. Mulch young plants with straw, with cottonseed or bone meal added. Remove suckers from the roots several times each year. Aphids and mites can be injurious. If weevils invade, destroy infested nuts. **Propagate** by layering. Suckers with roots attached may be planted in winter.

FIRETHORN, *Pyracantha coccinea*

Origin: Asia.

An evergreen, thorny firethorn is cultivated especially for its year-round beauty. This rounded bush grows to 10 feet in height and has glossy, dark green foliage. When set close together, the shrubs serve as a barrier hedge or a screen along a fence. It can be handled in a tub for a patio specimen if trained as a single standard. There are many varieties of firethorn, but the hardiest in cold-winter areas is Lalandi. Firethorns bear a profusion of small, clustered flowers on spurs. Fragrant and creamy-white, they are often visited by bees. From spring flowers, sprays of red-orange berries develop and ripen in the fall. With stems attached, this fruit may be boiled down to make an excellent jelly or jam. Add grapefruit juice to enhance flavor. A sauce can also be made from the ripe berries adding zest to cooked lamb, poultry, hamburgers and hot dogs. Sprays of berries brought in the house provide attractive bouquets, while berries left on these shrubs help to tide birds over rigorous winter months.

Culture: Firethorn grows fast and vigorously in most well-drained soils. It thrives in sunlight, and needs adequate water until established. Then keep moderately moist. Control growth by pruning. Firethorn is subject to aphids, red spider mites and scale. The worst disease in the West is fireblight, a virus that scorches and burns blossoms and tips of leaves. Prune away infested parts with disinfected tools and destroy to prevent spreading. **Propagate** by hardwood or softwood cuttings, by division of roots, or by layering.

GAYFEATHER, Blazing Star, *Liatris spp.*

Origin: North America.

Natives to the eastern and central United States, most species of these hardy perennial herbs are found in dry open woods, prairies, and along roadsides. Their flower heads are very showy when planted in an informal border. A thick tuberous radish-shaped rootstock produces the basal tufts of narrow, usually stiffish leaves. In summer, the tufts lengthen to tall, slender stems that are densely set with dotted, grassy leaves and topped by tiny rosy (sometimes white) fuzzy button-shaped flower heads. Tall gayfeather (*L. scariosa*) grows up to 3 feet. There are two noteworthy cultivated varieties: September Glory, with deep purple flowers, and White Sire, with white blossoms. Both grow to 6 feet high. *L. spicata*, gayfeather, grows to 6 feet high with 15-inch flower plumes. Two fine varieties are: Kobold, with deep rose-purple plumes, and 3-foot Silvertips, with lavender plumes. Blazing star (*L. squarrosa*), not over 2 feet high, resembles *L. spicata.*

The rootstocks of all these species contain medical properties. Cooked roots are home remedies for urinary disorders. A preparation made from the rootstock has been used as a sore throat gargle. The bruised rootstock may be helpful applied externally to snakebite, and a concoction of the roots in milk has been taken internally for snakebite.

Culture: Gayfeathers and blazing star tolerate any type of well-drained soil, but prefer it light. They need full sunshine and will endure drought. No fertilizer is necessary. Insects and diseases seldom endanger them. **Propagate** by dividing rootstocks in early spring. To prevent overcrowding, divide every three or four years.

GERANIUM, *Pelargonium spp.*

Origin: southern Africa.

Since their discovery in 1700, geraniums have undergone many changes through hybridization.   Most widely grown is *P. hortorum*, the upright zonals—so-called for the zones of color marked on the petioles inside the margins of their round velvety leaves.   These varieties are easy to grow, produce highly decorative blooms, adapt well to beds, pots, window boxes and planters.   Ivy geranium (*P. peltatum*), with its many varieties, has glossy, ivylike leaves.   Their long trailing stems make them graceful and popular basket plants.   Of most interest to herb gardeners are the *scented geraniums* filled with volatile aromatic oils.   Of the seventy-five species, about twenty-four have a distinctive scent and taste. General favorites are apple, cinnamon, lemon, mint, nutmeg and rose. Suitable varieties can be selected to impart flavor to jellies, custards, rice puddings, salad dressings and pound cakes.   Crushed leaves add flavor to fish, poultry, sauces and soups.   A scraped leaf placed on a small cut stops bleeding.   A tea made from the stems and leaves is an effective gargle for inflamed gums and to treat sore throat.   Simmered leaves added to bath water, act as a skin cleanser.   Flower petals may be used the same as other edible flowers.   Not hardy, scented geraniums should be treated as potted plants and brought indoors during winter.

Culture: Geraniums conform to average garden soil or potting mix. Good drainage is important; and soil must be allowed to dry out between waterings.   Provide full sun except for delicate types—such as

Lady Washingtons, dwarfs, and miniatures.   Encourage bushy growth and flower production by pinching.   Feed sparingly with bone meal when leaf color is poor.   Pests to watch out for are aphids, mealy bugs, red spider mites, thrips, whiteflies and slugs.   Leaf spot and stem rot can cause problems. **Propagate** in spring or fall from cuttings of firm shoots, inserting them in water or moist soil.

GINGER, Wild Ginger, *Asarum canadense and Zingiber officinale*

Origin: North America and Asia.

Now almost extinct, wild ginger once covered ground throughout North America. Indians collected the aromatic rootstock to make an infusion to relieve heart pains. In spring, this low-spreading herb with heart-shaped leaves bears brownish-purple flowers. It makes a good dense groundcover in the dappled shady areas beneath shrubs or trees. As a substitute for commercial ginger, the fresh rootstock is best peeled and sliced thin or grated. Add to sauces, stews, salad dressings and oriental dishes. Ground ginger is used in gingerbread, cookies, and pies. Chewing the raw root soothes a sore throat and stimulates saliva flow. Candied roots are delicious. If the water and syrup is saved, it becomes a gas-expelling remedy and is also helpful for colds. Hot ginger tea promotes perspiration to cleanse the body system. Easier to obtain is *Zingiber officinale*, common ginger, native to tropical Asia. Food stores stock these fleshy rhizomes, and their uses are the same as those for wild ginger. In most regions this canelike, stemmed plant must remain indoors during winter, since it needs 70° to 80° F. year-round warmth. Kept warm and in good light, the root can be dug up in ten months, having produced 6 to 8 additions along its sides. Replant the largest root, then wash the rest of the roots and dry in a warm place. Store them in the crisper of the refrigerator to use when needed.

Culture: Ginger requires well-drained, humusy soil, plenty of moisture, and shade from direct sunlight. During the growing season, feed common ginger every two weeks with fish emulsion. This encourages larger rhizomes and leaf sprouting. **Propagate** by division of rootstocks (rhizomes) early in spring while the plants are dormant.

GINSENG, *Panax quinquefolium*

Origin: Canada, eastern and northwestern United States.

For centuries Asiatic ginseng (*P. Schin-seng*), in the shape of a man with arms and legs, has been used in China to cure ailments, restore physical vigor, and to prolong life. The erect, perennial herb once flourished in the United States in rich soil, shaded by woodsy hardwood trees. Demand for the roots in China and Europe depleted the wild supply, and ginseng is chiefly found today under cultivation. American ginseng (*Panax quinquefolium*) has a spindle-shaped or forked root 2 to 4 inches long. The plant grows 8 to 15 inches high, bearing five thin pointed leaflets with toothed margins. In July and August greenish-yellow flowers appear, soon followed by bright crimson-red edible berries. Ginseng grows from seed never permitted to dry out, and takes a year or more to germinate. After eight years, mature roots are dug up and cleaned. Bits of thoroughly dry roots may be eaten raw or brewed into a rather bitterish tasting tea. Many Americans find that ginseng produces beneficial effects for various disorders, including resistance to stress, physical and mental fatigue, and for relief from internal temperature changes. Ginseng is nonaddictive and causes no side effects.

Culture: Ginseng requires a moist, deep, well-drained, fairly light soil mixed with forest leaf mold. It must be grown under lathy shade such

as provided by a framed arbor of deciduous vines. Well-decayed hardwood leaves worked into the soil are a good fertilizer, and a thick mulch of hardwood leaves applied in spring and fall is advisable. Mice may attack seeds and roots. Too much shade and water can cause fungus diseases. For detailed instructions on how to **propagate** ginseng, send for *U. S. Farmers Bulletin No. 2201*, Superintendent of Documents, U. S. Government Printing Office, Washington, D. C. 20402. The stock number is 001-000-02779-4 and the cost is $.35.

GRAPE, *Vitis spp.*

Origin: North America.

Since prehistoric times, grapes have furnished humans with nourishing food. Their natural sugar content is easily digested into energy. The woody vines grow wild along streams and borders of lakes, in foothills, and occasionally on desert edges throughout the United States. Although wild grapes are smaller—with tougher skins than cultivated forms—they render sweet, juicy fruit in profusion during fall. They may be eaten raw picked from the vine or made into jelly, jams, pies and grape juice. The large leaves if gathered when full-sized but still tender; add a luscious flavor to fish, game birds and poultry that is wrapped in them and baked. The vigorous vines provide shade and ornament over arbors, decks, terraces or fences. A thousand or more cultivated varieties and their hybrids have been named from several species of wild grape. By selecting plants that ripen in succession, fresh grapes can be enjoyed from midsummer throughout fall. Superior crops are determined by the varieties suited to your climate.

Culture: American grapes, including all their varieties and hybrids, need deep, well-drained soil, supplemented with compost or highly organic matter. Moderate summer sunlight is best. Cold tolerance is well below 0° F. Water deeply in spring and early midsummer, but give little water when the fruit is ripening. An alfalfa hay mulch provides nitrogen and keeps weeds down. Feed a vine in early spring with ¼ to ½ pound of *ammonium sulphate*, spreading it around to reach feeder roots. Generally grapes have few enemies or diseases if vines are well-fed and well-pruned. **Propagate** by taking hardwood or softwood cuttings of healthy canes, or by layering.

GRECIAN LAUREL, Sweet Bay, Urn Plant, *Laurus nobilis*

Origin: Mediterranean Region.

This is the laurel of history, grown in Greece as a tall tree or found wild on warm slopes around the Mediterranean Sea. In this country the slow-growing evergreen herb is commonly used as a decorative tub plant for porches, patios and pools. Bay grows 4 to 10 feet high in a pot. It needs winter protection in most regions, and can become a houseplant in winter if the foliage is misted occasionally. The dark green foliage can be sheared into various forms. However, on female plants this limits the appearance of the small, yellowish flowers and dark purple edible berries. These leaves and berries may be eaten to stimulate the appetite. They may be made into a poultice with honey or syrup and placed on the chest for colds. Oil pressed from the leaves and berries has also been made into a salve for rheumatism and bruises. Bay leaves act as repellents for moths and fleas. The slightly bitter-tasting leaves contain a volatile oil which is held by drying. As a meat spice, dried bay has the same usages as California laurel. Fresh or dried leaves used sparingly in the cooking water of bland vegetables, such as carrots, add a pungent flavor.

Culture: Grecian laurel prefers rich, well-drained soil. In hot summer climates, it needs filtered light or afternoon shade. Little water is necessary when established, but the plant should not be allowed to dry out during the growing season. Feed monthly from April to July with a 5-10-5 fertilizer diluted to half the recommended strength. Watch for scale insects. **Propagate** from stem cuttings taken at any season, rooting them in damp sand and peat. Suckers root quickly. They may be propagated by layering shoots in the fall.

GROUND CHERRY, Husk Tomato, *Physalis pubescens*

Origin: North America.

The bushy ground cherry is a perennial herb found in moist to dry fields, waste places and open country in all parts of the United States, except Alaska. It spreads rapidly over the ground, seldom reaching more than 1½ feet in height. There are many species of wild ground cherry, but *P. pubescens* is the one usually cultivated in gardens for its edible fruit. Large, ovalish leaves are produced on grayish-haired stems, springing from long, creeping underground roots. In summer, pretty bell-shaped, creamy-yellow flowers appear in the leaf joints. Later the five petals expand, become papery, angled and ribbed, completely enclosing the pulpy and seedy fruit. When ripe, the papery husks drop to the ground. The yellow fruit, resembling a small tomato, is not always ripe so it is best to gather the husks and let them dry for several weeks. Fully ripened fruit should be soft and very sweet. After removing the husks, the raw fruits are appetizing as desserts or in salads. They can be pleasing applied in pies, made into jams or preserves, or dried in sugar and used like raisins. Ground cherries are good stewed with sliced tart apples, or cooked with lemon juice.

Culture: Ground cherry is grown the same way as the tomato. Because its roots become invasive, it is well to raise the bush in a five-gallon container and use a soil substitute (see page 32). Mix into the material a liquid 5-10-10 fertilizer, following the labeled directions. After that, while husks are developing, feed monthly with a weak solution of the fertilizer. Stop fertilizing when husks near maturity. Keep the soil moist and the plant in full sun. Destroy blister beetles, whiteflies, slugs, worms and other pests as soon as damage is noticed. **Propagate** by division of roots in early spring. Seeds also germinate quickly.

GUM PLANT, Gumweed, Resin-weed, *Grindelia spp.*

Origin: North America.

About thirty species of gum plant are found in dry, open spaces, on mountain slopes and along roadsides, mostly throughout the western United States. This bushy perennial herb, at a height of 1 to 3 feet, has a round, smooth stem and long, sharply-toothed, leathery, resinous leaves. From August to December, the top of the plant brightens with small, sunflowerlike yellow flower heads, borne on branching stems. The buds shine with sticky resin. Long before doctors recognized its value, California Indians and pioneers were treating asthma, bronchitis, colds and whooping cough with teas made from the bitter-tasting gum. By 1875, the plant was authentically introduced into American medicine. The dried leaves and flowers of *G. squarrosa* are used as a sedative, to relieve spasms and other disorders. Doses taken in large quantities are poisonous due to the fact that all gum plants absorb selenium from the soil. But freshly gathered leaves and full-blooming flowers may be simmered in a small amount of water for

fifteen minutes to brew an external wash for curing poison oak, and this fluid can be used for other types of blisters and rashes, as well as burns. Commercial preparations of gum plant for these purposes are similar. *G. latifolia* and *G. robusta* make attractive ornamentals. They can liven a bare spot, be interplanted with different natives, or used in place of lawn.

Culture: Gum plant is not fussy about soil, and thrives in a sunny spot that tends to be dry. Fertilizing is unnecessary. It is practically immune to insects and diseases. **Propagate** by stem cuttings at any time.

HAWTHORN, *Crataegus spp.*

Origin: North Temperate Zone.

Countless species of native hawthorn prosper across the United States in hills near streams and meadows. They are dense and thorny trees to 25 feet in height. The gray bark made into a tea serves as a tonic for the heart, while the small glossy leaves make a pleasant tea. In late spring the charming white or pink flower clusters are a spectacular show. A tea made from the flowers is a remedy for both nervous conditions and insomnia. Clusters of red or purple berries come on in the fall. Red haw (*C. mollis*) produces red, shining berries which make excellent jam, jelly or marmalade. A decoction of the crushed berries is an herbal cure for sore throat and diarrhea. Nutlets from the ripened berries can be roasted and used as coffee beans. For best results with hawthorns, choose those species found as natives in your area. However, Washington thorn (*C. phoenopyrum*) adapts well to most regions and is considered the most outstanding of all natives. It has slender thorns and pendulous clusters of red berries that spread their charm well into winter. Fall foliage turns to brilliant red and orange. Shrubby types of hawthorn can become barrier hedges. Tree types can be featured as patio trees, or used to accent the landscape.

Culture: Hawthorns will grow in almost any well-drained soil and like either full sun or sunlight with partial shade. Keep them on the dry side to prevent ungainly growth. Prune away excess growth when a plant is dormant. Remove suckers at any time. Hawthorns are subject to many pests; fireblight and rust are common diseases, depending upon the area. **Propagate** from root cuttings.

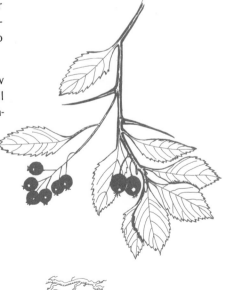

HONEYSUCKLE, *Lonicera spp.*

Origin: Northern Hemisphere.

These woody climbers or trailing shrubs can be found in moist woods and thickets throughout the United States. Out of the large number of wild species of honeysuckle, nearly all are worth cultivating for their brilliant flowers and ornamental berries. The usually deciduous, glossy foliage has been found to contain juices that kill common germs. Long before the coming of white settlers, Indians chewed the leaves and applied them to bee stings. They also smoked the nicotine-free dried herbal leaves as an aid for asthma. Orange honeysuckle (*L. cilosa*), a native of western mountains, climbs rampantly to the tops of tall trees. Its splendid clusters of scentless scarlet and gold tubular flowers are followed by bright red, edible berries. Yellow honeysuckle (*L. flava*) is one of the most delightful of the American species. A twining vine to 10 feet, this eastern native has fragrant reddish-orange flowers. Twinberry (*L. involucrata*), another western native, is an upright shrub 3 feet or more high. It bears scentless flowers of a dullish yellow, later displaying pairs of pea-sized shiny black berries. Trumpet honeysuckle (*L. sempervirens*) is a hardy vine well distributed across the country. Its unscented clustered flowers are bright red outside, with yellow interiors. Berries are red to scarlet.

Culture: Wild honeysuckles are all easily grown, thriving best in moist, loamy soil. Give them full sun, or light shade where summers are hot. Vigorous plants, they require pruning to control growth. Vining types need strong support. Their worst enemies are aphids. **Propagate** from stem cuttings taken from new growth.

HOREHOUND, *Marrubium vulgare*

Origin: Eurasia.

Horehound, a perennial herb, was brought here by early colonists for medical purposes, but has widely escaped. It is especially plentiful in California, growing along roadsides, in vacant lots, pastures, and by old buildings. The fiberous, spindle-shaped rootstock of the horehound sends up whitish woolly stems that spread into a bush about 2 feet high. Whorls of white flowers appear from the upper leaf axils of the fuzzy, gray-green leaves. The juice from the leaves is a popular, bitter-tasting flavoring used in cough drops and lozenges to soothe hoarse or irritated throats. A rich, volatile oil is retained in the stems and leaves just before the flowers open. The leaves are strong-flavored and should be used sparingly in salads, sauces and stews. Both leaves and flowers may be used, fresh or dried, to make a strong tea which is effective in combating colds, or for expelling intestinal worms. When made into a syrup with honey, the tea becomes a treatment for bronchitis and coughs. Leaves, stems and flowers are boiled down into what is known as horehound candy. Either the tea or crushed leaves may be applied externally to treat minor skin problems. Fresh and dried foliage is attractive in flower arrangements. The small seed pods collected during fall and winter may also be useful in craftwork. Horehound grows well as a garden plant among other herbs.

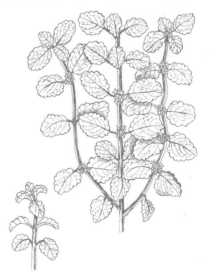

Culture: Horehound adjusts to any climate and favors poor, sandy, or light soil, kept fairly dry. Give it at least four hours of sun daily. Stake and trim to prevent sprawling. Like many herbs, horehound is disease and pest free. **Propagate** by root division, or by taking cuttings during spring or summer.

HORSERADISH, *Armoracia rusticana*

Origin: Eurasia.

Horseradish, a perennial herb, has been grown for centuries for its long, white, parsniplike root. It has become a weedy escapeé in cool, moist places throughout North America. Arising from the thick root are leaves of various shapes, differing from plant to plant. Leaves cooked with other greens, such as spinach, are a source of a wide variety of vitamins, particularly if the concentrated juice is reserved as a beverage. Roughly chopped young leaves are tasty in salads. Fresh horseradish root is strong in vitamin C, and can be stored in a refrigerator for months. It can thus be grated when required to release the flavor. Peeled and grated root is an ingredient for sauces, dips and salad dressings, stimulating the appetite and aiding digestion. Combined with white vinegar or lemon juice, the sharp taste is reduced. Freshly grated root is also helpful internally to loosen phlegm, relieve sinus-congestion, and to purify the blood. Diabetics can use horseradish safely as a seasoning for meat and fish.

Culture: Horseradish requires deeply dug soil for its long taproot. Mix in some coarse sand and compost if the soil is heavy. Choose a site at the far end of the garden because roots become invasive. Cool, damp climate is preferred, with plenty of water and sunshine. Mulch around the crown with compost during summer. Start harvesting in the fall, digging out roots as needed. Snails and pests of the cabbage family may strip the leaves; watch for them. **Propagate** in early spring from cuttings taken from side roots, slicing larger roots lengthwise. Cut tops straight across and the bottoms at an angle. Or, take a root with the crown attached, splitting the root lengthwise in strips, with a piece of crown attached to each one. This is the quicker method, and desirable where seasons are short.

HOUSELEEK, Hen and Chickens, Thunder Plant
Sempervivum tectorum

Origin: Europe.

This common but attractive succulent 3 to 8 inches high, has long been cultivated in Europe and is often found growing wild on dry mountain slopes, in stony soils, on walls, even on rooftops. On European housetops, houseleek is believed to act as a natural lightning rod to reduce the chance of lightning striking. Here the perennial grows well in rock gardens, containers, encrusted in crevices, pavement and boulder pockets, or wherever a suggestion of soil is present. The fleshy, pointed, gray-green leaves huddle together in a packed, globe-shaped rosette to survive the glare of the sun. Each wedge-shaped leaf contains a reserve supply of juice to enable it to overcome the most adverse conditions. Springing from the dense rosettes are hairy flowering stalks about 12 inches high, dividing into 1-sided clusters of small, pinkish-red, starlike flowers. *S. t. triste* is a form of houseleek with reddish foliage. A poultice of the freshly bruised leaves is a safe external remedy for headache, insect bites, inflammation and swelling. Pressed leaves, discharging much liquid, can be used as an antiseptic on cuts and burns. Fresh juice from a leaf has been employed as a wash for inflamed eyes, as well as applied to sties, warts, corns and freckles. Internally, an infusion of the leaves can be taken for shingles or worms.

Culture: Houseleek does not like rich soil nor much moisture. To protect against standing ice and water during winter, try to plant it where water never collects in pools. Sun and drainage is important. Watch for mealy bugs, slugs, snails and sowbugs. Root and stem rot are caused by over-watering. **Propagate** by leaf cuttings inserted in sand, or by removal of offsets in the spring.

HYSSOP, Blue Hyssop, *Hyssopus officinalis*

Origin: Eurasia.

At least as far back as Biblical times, the somewhat woody herb hyssop has been grown for medical purposes and for seasoning food. Since its introduction to this country from southern Europe, this hardy, bushy plant has become naturalized in sunny locations and on mountain slopes. The small, narrow, shining, dark green leaves are attractive and strongly scented. Growth is about 2 feet in height. Deep blue-lipped flowers and their varieties of pink, purple or white, bloom at the tops of the branches and stems from July to November. Hyssop flowers attract the destructive cabbage moth which feeds on Brussels sprouts, cabbage, kale and kohlrabi. Planted in the vegetable garden, hyssop in flower serves as an insect deterrent, and lures bees. Its roots deter many soil pests. The plant improves production of grapes cultivated close-by. A few of the bitter, slightly minty-flavored leaves add a pleasant taste to raw salads and fruit cocktails. Tender young flowering stems can be used fresh or dried for seasoning broths, soups, sauces and stews. A strong tea made from the leaves and flowering tops aids poor digestion, and relieves coughs, throat and nose congestions. Tea taken with honey cleanses the intestines. A poultice of

crushed leaves mixed with sugar is a remedy for avoiding infection of deep cuts. Leaves steeped in hot water remove discoloration caused by bruises.

Culture: Hyssop likes rather sandy soil and good drainage; it must have sunshine to produce flowers. Tops should be cut back frequently to keep young leaves growing. In moderate climates, hyssop remains an evergreen the year round and lasts for years with minimum care. **Propagate** by stem cuttings, or by division of mature plants. Within six weeks a root division will yield a crop of fresh leaves.

JAPANESE ARALIA, *Fatsia japonica*

Origin: Japan.

A lovely decorative evergreen shrub or small tree, this import is widely planted for its interesting, tropical foliage. In severe winter climates, Japanese aralia must be planted in a container and placed indoors by a well-lighted window. The glossy, dark green, fanlike leaves grow up to 16 inches wide on long stalks. Of moderate growth, the plant itself rarely reaches more than 8 feet in height. Pruned from the top in early spring, it can be kept to 4 feet. This yields a bushier plant and encourages side growth. In favorable situations, Japanese aralia produces delicate, small, creamy panicles of flowers in winter which are followed in spring by clusters of shiny black unedible berries. For larger and hardier leaves, snip or pinch off the flower buds when they appear. This exotic plant is best used as a single specimen to display its bold beauty on a shaded patio, porch, or entryway. It makes a good year-round houseplant anywhere. Very young plants are effective in a terrarium. Variety Moseri grows low and compact, while variety Variegata has leaves irregularly edged with golden yellow or creamy white.

Culture: Japanese aralia grows well in nearly all soils. For containers, use equal parts of loam, sand, peat, and leaf mold. Give it full shade, but in cool summer climates it will tolerate some sun. Keep the plant evenly moist and spray with a hose frequently during hot weather. Feed an established plant monthly with a diluted solution of liquid fertilizer. Aphids, mealy bugs, red spider mites, scale and thrips may attack. Destroy slugs and snails. Yellow foliage indicates too much sun. **Propagate** by stem cuttings taken from large plants in the spring, or by severing suckers that arise from the base of plants.

JASMINE, Common White Jasmine, *Jasminum officinale*

Origin: northern India and Persia.

The common white jasmine is one of the best known and most highly prized of the hardy cultivated woody climbers. A partially evergreen to deciduous vine, it grows to 30 feet or more if not pruned back severely and shaped to control its natural luxuriant growth. It can be planted directly in the ground in warm regions. Elsewhere, this twiner must be confined in a large container and returned to the house before the first frost. In either case, a support such as a trellis or pole is required for the spreading, clinging branches. The plant is used in India for snakebite, and the dark green, glossy leaves are a remedy for eye ailments. Beautiful sweet-scented flowers bloom from June to October. Waxy, white and funnel-shaped, they open up in many flowered clusters at the branch tips. Oil extracted from the flowers is made into a pleasant-smelling perfume. Centuries ago Persians used jasmine oil to scent their banquet rooms. Today the flower petals are a common ingredient for potpourris. A fragrant and soothing tea may be made from the petals.

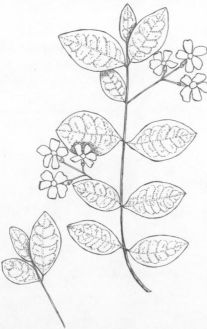

Culture: Jasmine thrives in the ground in ordinary garden soil. If potted, use equal parts of loam, sand, and peat. Provide bright light but not direct summer sunlight. Let the soil dry out between waterings and give less water during fall and winter. Feed every other month in spring, summer and fall with a 5-10-5 fertilizer, withholding food in winter. Enemies to jasmine are mealy bugs and scale. Should black sooty mold appear, look for mealy bug infestations. **Propagate** by softwood cuttings in spring, by semihardwood cuttings in late summer, or by layering.

JERUSALEM ARTICHOKE, Sunchoke, *Helianthus tuberosus*

Origin: eastern North America.

The name Jerusalem artichoke is a corruption of the word *girasole*, the name given the vegetable in southern Europe after early explorers learned from the American Indians the value of the tubers and took some home with them. Extensively used in colonial gardens, the plant has escaped to moist, sandy soil along ditches, fields and streams throughout the eastern United States. This vigorous perennial grows 6 to 10 feet high, and has the look of a sunflower. Unopened bud heads can be boiled and eaten like globe artichokes. The bright yellow flower heads open in late summer, seldom forming seed. These can be chewed, but it is for the nutritious tubers that the plant is grown. Attached to creeping roots, they are dug after the first frost. Peeled, sliced and eaten raw in salads, the tubers offer the crispness of water chestnuts and the taste of ripe coconuts. They are also delicious when boiled or baked. The jellylike substance that develops after boiling offers an appetizing base for soups. Because the tubers lack starch, they are excellent for those on a diet or those who need easily digested food. The non-mealy tubers may be used in place of potatoes in numerous recipes which call for the more common vegetable. A few plants in the back of the garden ensure an unending supply of these vitamin-packed tubers.

Culture: The hardy Jerusalem artichoke appreciates moisture, sunlight and soil enriched with fertilizer. For the best flavor dig tubers within twenty-four hours of eating. Few if any insects attack the plant. **Propagate** by planting whole tubers in early spring, or by cutting "eyes" the same way as potatoes.

JERUSALEM CHERRY, Christmas Pepper, *Solanum spp.,* *Capsicum spp.*

Origin: Madeira and South America.

Jerusalem cherry (*S. pseudo-capsicum*) is a popular Christmas gift, frequently living from year to year under favorable conditions. It forms into a small bush of glossy, tender, evergreen leaves. The abundance of white, star-shaped flowers are replaced by globular green fruits that eventually turn brilliant orange, red or yellow, often measuring an inch or more in diameter. These attractive "cherries" last a long time and are to be admired, not eaten. Although tempting to young children, the fruits contain poisonous juices. Tall varieties of the plant should be pinched and pruned to control growth. Closely related is Christmas pepper (*Capsicum annuum*) and its varieties make colorful pot plants. The starry white flowers are followed by small fruits of different sizes and shapes, usually flaming scarlet when ripe. Fruits are edible, but very hot. Both of these plants profit from summering outdoors in a shaded location.

Culture: These plants prefer equal parts of loam, sand and peat for soil. They grow well indoors in a sunny east, west or south window. Water thoroughly when the soil surface starts to dry. Mist the plants daily if the atmosphere is dry. A house temperature of not over 75° F. is best. Feed monthly with a weak solution of 5-10-5 fertilizer during the growing season. Aphids, scale, thrips and whiteflies like to suck on the tender new growth. Destroy them as soon as detected. Leaf and fruit drop are caused by poor light or overwatering. **Propagate** from stem cuttings taken from old

plants in March or April, rooting them in sand and in a warm shady place. To renew an old plant, cut off about two-thirds of the top growth at the end of the fruiting season. Then take it out of the pot and remove most of the soil from the roots. Repot in fresh, moist soil.

LAMB'S EARS, Betony, Chinese Artichoke, *Stachys spp.*

Origin: distributed world-wide.

Lamb's ears (*S. olympica*) is a woodsy perennial herb, growing not over 18 inches high. Where it has escaped, it is usually found in shaded, wet places. The silver-white, soft foliage changes with the light and resembles woolly ears. With its evergreen and dense leaves, the plant is an asset all year to contrast with dark green foliage in borders and edgings. It is also desirable as a groundcover beneath a garden tree or under shrubs. Pinkish-purple flowers appear in terminal spikes from June to August and are readily pollinated by bees. Dried leaves made a pleasant tea, while the dried flower spikes add to flower arrangements. Similar is betony (*S. officinalis*), which grows to 24 inches in height. It has more or less fuzzy leaves and bears reddish-purple spiked flowers in summer. A former medical plant, a tea made from the leaves is a home remedy for heartburn, nervous tension and sweating. Mixed with other herbals, the dried leaves have been smoked to counteract asthma, bronchitis and other pulminary problems. Juice extracted from the plant may be used externally to heal cuts and sores. Chinese artichoke (*S. sieboldi*) is a vegetable with slightly hairy leaves that grows to 18 inches high and produces edible tubers. The small spikes of flowers are white or pink. In the fall, the small and numerous tubers are dug as needed, to be eaten raw, boiled or roasted.

Culture: All three of these species grow best in well-drained humus. Plant them in shady spots with damp conditions. These herbs exude a rather musty odor that tends to repel insect pests. They are practically disease free. **Propagate** the first species by division of their fibrous roots in September, March or April. Divide and plant tubers of Chinese artichoke in April.

LAVENDER, English Lavender, *Lavandula spica*

Origin: Mediterranean Region.

In its original home, this classic lavender thrives in stony, dry semiarid areas. It was a popular bath perfume of the Romans and is still a basis of many herbal products. This woody, sweet-scented perennial rises to a height of 1 to 3 feet and has many narrow white felty leaves which give the plant an ash-gray semblance. It will flourish in a rock garden or as a border to a sunny garden path. Lavender can also be grown in containers. Where cold is extreme, it must be wintered indoors. The pale lavender flowers appear in spiked clusters, are fragrant, and attract bees. Although the fresh flowers are the most aromatic, the whole plant contains an essential oil that can be saved in drying. Flowering tops should be removed just before they bloom. Dried leaves burned as incense, suppress kitchen odors. Infusions of the flowers and leaves have been used for dizziness, fainting, nausea and vomiting. Buds cut before they open are useful for food flavoring and potpourri. Dried flowers and seeds placed in cheesecloth bags can be used in bath water to soothe nerves. If placed in drawers or hung in clothes closets among woolens, the pleasing aroma of lavender deters moths.

Culture: Lavender likes loose, fast-draining soil in a sunny location. It resists drought and needs no fertilizer. Snip off flower buds the first year to produce better blooms the second year. Thereafter, remove faded flowers to keep the plant blooming. Insects and diseases are not a problem. **Propagate** by root cuttings in winter, and set out in spring; or, from stem cuttings taken in summer with heels attached. It is necessary to replace lavender every seven years because after that plants dry out.

LEMON BALM, *Melissa officinalis*

Origin: Mediterranean Region.

This delightfully sweet-scented perennial herb receives its name for its distinct lemony odor. It has become naturalized in some parts of the United States and is found in partially shaded, moist fields, old gardens, and along roadsides, growing 18 to 24 inches tall. The bushy plant has crinkly, pear-shaped, heavily-veined leaves, with clusters of smallish, yellow-white flowers that shoot forth from short leaf stems. These late summer blooming flowers furnish pollen to bees, which in turn produce a lemon-flavored honey. Raw lemon balm leaves give a lift to vegetable and fruit salads. The crushed aromatic leaves are useful for polishing furniture, or for making a poultice to abate the pain of insect bites. Bruised leaves, fresh or dried, emit sufficient scent to perfume a room. Freshly ground leaves add a lemon-mint flavor to beef dishes, fish, or lamb, or may be used as a substitute for lemon in tea. Lemon balm tea made from the dried leaves is a remedy for nervous indigestion, migraine headaches, feverish colds and insomnia. Cut branches keep well in flower arrangements. The plant spreads rapidly and is best confined in a large pot where it can not become a pest. Potted in the ground among other herbs, the yellow-green foliage makes a pleasing contrast. In most climates, the clump dies back in winter unless brought indoors and kept on a window sill.

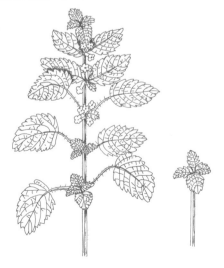

Culture: Lemon balm grows best in rich, moist soil, and enjoys partial shade. Mealy bugs, mites and whiteflies may be bothersome. Trim the plant occasionally to maintain compactness. It should live for years. **Propagate** by root division, by layering, or from cuttings taken from new growth in late summer or early fall. Upper portions of old stems may also be used for cuttings.

LIVERLEAF, Liverwort, Mayflower, *Hepatica spp.*

Origin: eastern North America.

These charming perennial herbs, usually under 6 inches high, grow wild on moist wooded hillsides throughout the eastern United States, Canada, and south to Florida. They will thrive in any part of the country if given proper care. These hardy and adaptable plants are attractive in a shady rock garden, or grown in pots. The many-branched rootstock produces a rosette of 3-lobed leaves which are green on top and reddish-purple beneath. *H. acutiloba* has acute-lobed leaves, whereas *H. americana* has blunt lobes which are usually broader than long. The leaves thicken and last all winter, with soft and furry new leaves appearing after the plant blooms. Brilliant blue-violet, purple, white or rose-pink flowers flourish on hairy stalks from December until May. In extremely cold climes they blossom on a bed of snow. Following the flowers come silky fruits. Leaf shapes resemble the human liver and because of this the plants were once classified as medical herbs. The leaves were used in a tea to treat hepatitis, liver congestion, gall bladder problems and other ailments related to the kidneys and bladder. Hepatica tea, made from the fresh or dried leaves, has been prescribed by modern herbalists for bronchitis and to stop internal hemorrhaging. Large doses can produce symptoms of poisoning. Therefore, the tea should be taken with caution.

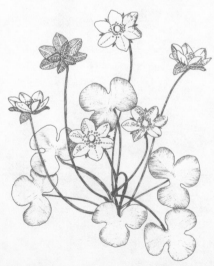

Culture: Liverleaf requires damp acidic woods soil, full shade and good drainage. Although not difficult to transplant, it takes time to become established. Rust fungus can be a problem. **Propagate** by division of rootstocks after the flowering season.

LOVAGE, Sea Parsley, *Levisticum officinale*

Origin: southern Europe.

This huge perennial herb is found wild along the North Atlantic coast and is believed to have escaped from colonial gardens, where lovage was popular for cooking and medical purposes. A short, thick rootstock sends up round, branched, hollow stems 3 to 7 feet high. Leaves are shiny and brilliant green, looking like celery and similar in taste. Used as a potherb, lovage renders a yeastlike flavor. Tender young leaves, fresh or dried, offer a good substitute for celery in raw salads, sauces, soups and stews. Dried and powdered leaves can serve as a salt substitute. Fresh stems may be blanched and eaten raw like celery, or sliced and added to soups. A tea or broth made from the stems and leaves stimulates the kidneys and is helpful for rheumatic problems and digestive upsets. In summer, small pale yellow flowers grow in compound clusters. Seeds from the flowers are delicious crushed or whole in candies, cakes, salads, soups, meat pies or roasts. The pungent and aromatic roots give a celerylike taste to herb vinegar and vegetables such as string beans and limas. Roots and stems of lovage may be candied. It grows rampant when mature, and one plant of the attractive herb is sufficient for a family. Plant in the garden or in a tub. It will give a dramatic, colorful effect to any patio.

Culture: Lovage grows best in rich, deep soil with good drainage. Give it full sun or partial shade. Feed the spreading roots in spring with compost or manure. To produce larger leaves and stems the first season, remove flowering stalks before they develop. Control aphids, caterpillars, slugs and snails that invade young leaves. **Propagate** by dividing the roots in spring or fall, following the second or third year of growth.

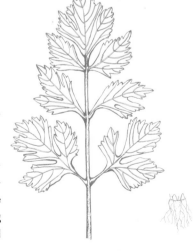

MARJORAM, Sweet Marjoram, *Marjorana hortensis (Origanum marjorana)*

Origin: southern Europe.

Marjoram is native to sunny places along the shores of the Mediterranean Sea. A perennial herb, it has been cultivated since ancient times for its culinary and medical value. Where winters are freezing, marjoram must be carried over in a sunny window sill. Elsewhere, grow it in a container near the kitchen door to be readily available. The plant forms a compact clump 12 to 18 inches high, with gray-green oval leaves. In midsummer, tiny white starlike flowers appear in clusters at the top of soft downy stems. When used internally and regularly, the powerful antiseptic herb helps the body to resist diseases. It also has a beneficial effect on gastritis and indigestion. Chewing the leaves will deaden a toothache and sweeten the breath. Applied externally, oil from the aromatic leaves aids stiffening joints. A delicate, perfumed tea made from the leaves has a calming effect and relieves nervous headaches. Weak tea is used for colic in children. In a sleep pillow, bruised leaves tend to induce slumber. The fragrant leaves are delightful mixed in potpourris. In cooking, the fresh or dried leaves add pungency to tomato soups, poultry stuffings, stews, vegetable water, omelets, gravies and meat loafs. Finely chopped marjoram leaves improve the taste of green salads or their dressings. Start cutting the leaves when the plant is 4 or 5 inches high to prevent blossoming. Dry the excess leaves, then crush them to store in airtight jars for later use.

Culture: Marjoram likes well-drained, fairly moist, light soil. If given a sunny and sheltered spot, the plant will last for years. Dispose of mealy bugs, mites and whiteflies. **Propagate** by stem cuttings in the spring, root division, or by layering.

MANZANITA, *Arctostaphylos spp.*

Origin: Pacific Coast.

Across the mountains from California to Alaska there are often heavy stands of evergreen manzanita, ranging in size from creepers to tall shrubs and small trees. Most admired features are the crooked branches with smooth red to chocolate-colored bark, handsome foliage and tiny applelike fruits. Many of the forty-three species adapt well to garden situations, but all are important bee plants. Indians and early settlers crushed the leaves for a poultice to heal sores. In a fluid extract, the leaves were a treatment for poison oak. The extract taken internally aided bladder and kidney disorders, stomach problems and headaches. A pleasant tea is brewed from the waxy, white or pink bell-shaped flowers. The sour green berries are thirst quenching, and can be made into a soft drink or an excellent jelly. When fully ripe, the fruit pulp is dry and sometimes sweetish. Ripe berries may be eaten from the bush, dried, or stewed. They are good ground into meal after removing the seeds. Boiled berries make a tasty drink if sweetened with honey. Unboiled manzanita cider conserves the rich vitamins in the berries. Hairy manzanita (*A. columbiana*) grows strongly branched from 3 to 15 feet tall and is widely cultivated. The white flowers are followed by red berries. *A. manzanita*, common manzanita, will reach 20 feet in height; it produces white to pink clusters of flowers, with the fruit turning from white to deep red. *A. uva-ursi* (bearberry, kinnikinnick) is a hardy groundcover that roots itself as it spreads and creeps. Its glossy, leathery leaves become bronzed in winter. Flowers are white or pink, berries red or pinkish.

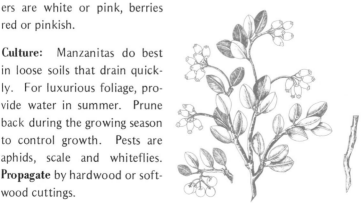

Culture: Manzanitas do best in loose soils that drain quickly. For luxurious foliage, provide water in summer. Prune back during the growing season to control growth. Pests are aphids, scale and whiteflies. **Propagate** by hardwood or softwood cuttings.

MINT, *Mentha spp.*

Origin: Eurasia.

There are some forty species of strong-scented mints, including their varieties. Many have become naturalized in the United States and are found in damp, partially shaded areas. All are hardy perennial herbs 12 to 24 inches high, with lavender to purple spiked flower heads. Each differ in strength and flavor, but most can replace each other for seasoning or medicine. Since mints spread rapidly, they are best confined in containers and placed in the garden where they can deter destructive insects. Fresh mints contain a high content of vitamins A and C. They are used to cure night blindness and scurvy. Fresh leaf mint tea is an old remedy for chills, colds, upset stomach and hiccoughs. The fresh leaves are also used to make delicious jam and jelly. Mints impart flavor to lamb, veal, cream cheese, tossed salads and vegetables. The common field mint (*M. arvensis*) makes especially fine jelly and flavorsome hot or cold drinks. Orange mint (*M. citrata*) has a citrus flavor that enhances fruit salads and punches. Peppermint (*M. piperita*) is a soothing digestive aid in candies, chewing gum and medicines. It is also helpful in discouraging flies, mice, and mites from entering buildings. Pennyroyal (*M. pulegium*) has a strong peppermint flavor. It is a good condiment in soups and stews. Apple mint (*M. rotundifolia*) produces woolly gray-green leaves that emit a sharp, sweet perfume when chopped. Spearmint (*M. spicata*) is the most commonly cultivated mint and holds the greatest amount of aroma. The content of menthol oil in the leaves makes it effective as a tea to remedy diarrhea and neuralgia. Also, it efficiently drives away aphids.

Culture: Mints thrive in soil provided with organic matter. They like the sun in the morning, and plenty of moisture. Snip off flowering stems to preserve the sharp, mint flavor. Burn plants affected with rust. **Propagate** by root division in the spring, by cuttings, or layering.

MOUNTAIN-MAHOGANY, *Cercocarpus spp.*

Origin: western United States.

These remarkable evergreen or deciduous shrubs and small trees struggle on dry, rocky slopes of foothills and mountain ranges throughout the West. Although called mountain-mahogany, they are not related to true mahogany. Out of twenty known species, two are commonly cultivated in gardens. Their open crowns and branching patterns, lush foliage, and furrowed bark (found on older plants), give them a distinctive charm. Silky clusters of white flowers cover the crowns in spring, sweetening the air for about two months. They triumph in summer and fall when the long-lasting small fruits appear, topped by long, twisted, feathery plumes that glisten in sunlight. Hardtack, or mountain ironwood, (*C. betuloides*) becomes a tall shrub, but can gradually form an interesting scraggly tree to 20 feet in height. The veiny, deciduous, curled leaves are gray-green above the pale beneath. Navaho Indians used the streaming plumes for prayer sticks, and the hard wood to make dice. Curl-leaf mountain-mahogany (*C. ledifolius*) is a twisted and curled tree which eventually reaches 20 feet high. Its evergreen, leathery, resinous leaves, rolled under at the edges, are dark green above, hairy and white beneath. When taken over a long period, a decoction made from its bark has proven successful in curing venereal disease. Indians of the Rocky Mountain regions scraped, dried and boiled the inner bark, from which they made a tea to cure colds, pneumonia, and to regulate the heart. Indians also pulverized the reddish-brown outer bark to apply to serious wounds and burns. California Indians used the extremely hard wood for fuel and to make clubs.

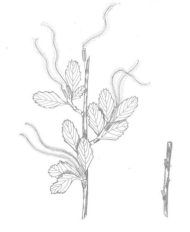

Culture: Mountain-mahogany needs deep, well-drained soil and open sunlight. Drought-tolerant, it grows well with little or no water once established. **Propagate** by hardwood or softwood cuttings.

NIGHT-BLOOMING CEREUS, *Epiphyllum oxypetalum*

Origin: Mexico to Brazil.

In the wild, these fantastic cacti flourish in damp rain forests where most perch upon debris or moss in forks of tree branches. The green, jointed stems are flat, smooth, usually scalloped along the edges and resemble chains of leaves. Night-blooming cereus are cultivated chiefly for their magnificent flowers that bloom from April through June on the previous season's growth. Although rangy and gawky most of the year, the joy of discovering flower buds and watching them burst into gorgeous white flowers at night certainly makes growing this plant worthwhile. Hybridization has produced blossoms as much as 10 inches across, vying in size not only with the largest cacti but with all flowering plants. Other varieties bear flowers in such profusion that not even orchids can rival them. In addition, there are numerous day-blooming hybrids called orchid cacti. All *Epiphyllum* are normally large, strong and spineless plants, but some of the modern hybrids are smaller in growth. Among the thousands of named kinds, nearly every size, shape, and flower color can be found. Many of the flowers show more than one blend. *Epiphyllum* need winter protection, such as a sunny east or west window. During summer, they do best outdoors under the shade of trees or in a lathhouse. Due to their long, trailing stems tubs or hanging baskets are suitable containers.

Culture: Night-blooming cereus like equal parts of loam, sand and leaf mold for soil. Indoors they

prefer a temperature not exceeding 75° F. Keep them evenly moist except in winter when they need only enough water to keep the stems from shrinking. Feed with a 10-10-5 formula fertilizer biweekly in spring and summer, using it at half strength. Watch for attacks from aphids, mealy bugs, snails and slugs. **Propagate** by rooting stem cuttings in spring and summer.

OLIVE, *Olea europaea*

Origin: Mediterranean Region.

The long-lifed olive tree has branches extending slowly to 25 feet or more in height. Since prehistoric times a native of hot, dry areas, it now grows well in parts of this country where the climate is similar. Elsewhere it can be put in a tub and wintered indoors. Evergreen, soft gray foliage complements a smooth gray trunk which becomes gnarled with age. A medicine made from the leaves or yellow inner bark reduces fever, while an infusion of the leaves acts as a tranquilizer to help relieve nervous tension. In late spring fragrant clusters of white flowers open, to be followed from fall to early winter by quantities of shiny purple to black fruit. Oil pressed from the ripe, processed fruit has value as a laxative. It soothes the mucus membranes, and is regarded by some to desolve cholesterol in the walls of the arteries. With the addition of alcohol, olive oil makes an excellent hair tonic. It is used as a base for liniments and ointments. The oil retains its natural flavor in cooking or salad dressings. Ripe olives are often used as appetizing ingredients in hot dishes, salads, sauces and sandwich spreads. A number of olive varieties are available, each differing in size, oil content and flavor. Most are self-fruiting, though there are fruitless varieties.

Culture: Olives produce best in deep, rich soil, but tolerate almost any soil that is well-drained and moderately dry. Full sunlight is important. Feed mature plants yearly with one or two pounds of nitrogen. Thin out branches each year to reveal the beauty of branch patterns. Watch out for scale insects. To prevent olive knot from spreading, remove diseased parts as soon as swellings show. **Propagate** from cuttings of hardwood, semihardwood or softwood. Other means are layering, root cuttings and suckers.

PEACH AND NECTARINE, *Prunus Persica*

Origin: China.

One of the first trees brought to this country was the hardy and fruitful peach. Many have escaped and grow without help in meadows, mountains, or around abandoned farms. There are few places unsuited for some type of peach. Today hundreds of named varieties flourish, differing mostly in climate adaptation. Nectarine fruits lack the fuzz of the peach though trees look the same. Both bear yellow or white fleshed fruits, either clingstone or freestone. Yellow fleshed fruit is richest in vitamin A and the content is not removed by cooking, canning or freezing. Peach and nectarine trees come in dwarf and standard sizes. Standards grow rapidly to 25 feet, however, their growth can be easily controlled by pruning. An abundance of blossoms appear before the leaves in spring, ranging from a pale pink color to dark red. Most are self-pollinating. The fruit generally ripens in midsummer and is picked when the color is yellow, or white in "white" varieties. To avoid bruising, peaches must be handled with care. Aside from the juicy, delicious fruits, the thin leaves have been used as a mild sedative for nervous conditions, as a laxative, and to stimulate the flow of urine. A hot tea made from the leaves and given in small doses is said to stop vomiting. Powdered dry leaves and bark have been applied to sores and wounds to hasten healing.

Culture: Peach and nectarine trees do best in a loamy soil well mixed with organic matter. Good drainage is essential. Plant them in a sunny

site not subject to winter freezing. Feed around the root zones in late fall with organic fertilizer. After fruiting, prune heavily to increase vigor; train to form. Serious insect pests are scales and borers. Worst diseases are bacterial leaf spot and a contagious virus called "yellows." **Propagate** by softwood cuttings made from young spring growth and started in a covered propagation box.

PERSIMMON, *Diospyros virginiana*

Origin: United States.

The deciduous persimmon tree is found wild in most parts of this country, growing in dry fields, open woods or clearings. Spring flowers are followed by plumlike green berries which remain puckery to the taste until frost or complete ripening changes them into beautifully colored yellow to yellowish-red, soft, edible fruits about 2 inches across. Dried roots of the 20- to 30-foot high tree have been used as a tonic. The attractive gray-brown, checkered outer bark is highly astringent. When boiled into a dark liquid, it becomes a bitter-tasting remedy for diarrhea, dysentery, and uterine hemorrhages. Although bitter, the boiled inner bark is a good agent for fevers and sore throats, while the astringent unripe fruit mixed with alum is an effective gargle for ulcerated throats and mouth ulcers. The glossy leaves that turn lovely hews in the fall are highly nutritious and can be taken to prevent scurvy. Tender young leaves are fermented and dried, then used like black tea. Hard-ripe persimmons may be peeled, hung in the sun to dry and reserved for future use. When soft-ripe, the fruit loses its astringency and tastes sweet. The fruits are luscious raw in fruit salads, cooked in bread loafs, or made into jam and pudding. Several varieties of native persimmons are under cultivation, each abundantly bearing fine quality fruits. All make excellent shade trees in the garden and grow anywhere a peach will.

Culture: Persimmons thrive in any well-drained, fertile soil. The trees only require enough water to prevent the soil from drying out. A straw mulch should be applied in early spring. Prune to remove suckers, deadwood, and to shape a tree. Serious pests and diseases seldom attack. **Propagate** by softwood cuttings made from spring growth and started in a covered situation.

PINEAPPLE GUAVA, *Feijoa sellowiana*

Origin: South America.

Pineapple guava is among the hardiest of subtropical fruits and can resist temperatures down to 15° F. for short intervals. A shrub or small tree, it may reach a height of 15 feet if left unpruned. The many branches bear 2- to 3-inch-long leaves which are glossy green on top and chalky white beneath. This evergreen color effect produces a lovely sight in the garden or as a container subject. Exotic purplish-white small flowers bloom in May or June with bristly tufts of crimson stamens. The thick, sweet-tasting flower petals find use in fruit salads. Round or pear-shaped fruits become gray-green in late summer, falling to the ground when ripe. Inside, the fruit is filled with soft, white pulp and surrounded by a mass of seeds. The pineapple-flavored pulp becomes something-to-remember when served raw for dessert, or processed into jam, jelly, paste or juice. Frozen juice is also good. Most varieties of pineapple guava need cross pollination to bear fruit, but Coolidge and Pineapple Gem are exceptions. Strawberry guava (*Psidium cattleianum*), from tropical America, has small, dark-red fruits. The white pulp, with the tart taste of strawberries, is used in the same manner as pineapple guava. When trained as a 15-foot tree with a multiple trunk, the beautiful grayish-green to golden-brown bark is exposed. Enhancing the tree's worth are the glossy leaves and showy, long-stamened white flowers.

Culture: Both guava types do best in soil enriched with well-rotted

manure, but adapt to nearly any moist soil. They like good drainage and full sunlight. Mulch them heavily with hay or straw. Feed every six to eight weeks during the growing season with cottonseed meal or fish emulsion. Destroy whiteflies and scale insects that attack. A planting made in soil containing not fully decomposed roots can cause crown and root rot. **Propagate** by root cuttings.

PLANTAIN, *Plantago spp.*

Origin: world-wide.

Plantains are despised weeds in lawns, gardens and agricultural lands, inhabiting anyplace in the United States where there is moisture and sunlight. Although considered obnoxious pests, these low-growing perennial herbs are naturally nutritious and contain fine medical properties. Their young, shredded leaves put into green salads, vegetable soups, meat casseroles and custards, furnish a rich supply of minerals and vitamins A and C. Tender young leaves cooked lightly and seasoned like spinich, make a palatable vegetable. Flowers borne on long leafless stems turn into smooth, shiny seeds which can be ground into a wholesome meal for breadmaking. Seeds soaked in water can be eaten for a laxative. Buckhorn plantain (*P. lanceolate*) is a species introduced from Europe. When mashed into a paste and applied as a poultice, its hairy, narrow leaves have a cooling and soothing effect on boils, carbuncles, cuts, and hemorrhoids. Fresh juice pressed from the entire plant is helpful for throat irritations and hoarseness. More abundant is common plantain (*P. major*), with smooth broad ovalish leaves that are conspiciously ribbed. Its uses are the same as for buckhorn plantain. In addition, the juices have

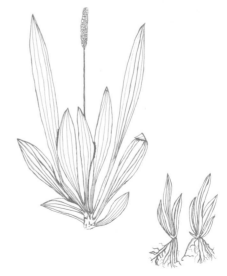

been used to remedy diarrhea, and kidney and bladder problems. The mashed green leaves have been applied to bruises, inflamed eyes, snakebites and bee stings.

Culture: Plantains grow in rich soil where there is plenty of water and sunshine. Plant them in large containers to prevent spreading to other parts of the garden. Remove flower heads not intended for use before seed capsules split open. **Propagate** by root division.

POMEGRANATE, *Punica Granatum*

Origin: southern Asia.

More than 4000 years ago the pomegranate was cultivated in Persia. Today it grows as a shrub or tree up to 10 feet tall in parts of the United States where summers are hot and dry. In cooler climates and where it becomes extremely cold in winter, the plant will still produce edible fruits when grown in a tub and sheltered indoors. It makes a lovely sight in the fall, with its glossy green foliage turning to brilliant yellow. The usually reddish-orange hibiscuslike flowers are followed in the fall by dark red, sometimes brownish, fruits. Upon ripening, the flesh inside the leathery rind is red, juicy and sweet-tasting. Although seedy, the flesh is delicious eaten out of hand or when served on fruit salads as a garnish. The syrup grenadine is made from pomegranate juice. From the finely ground rind an excellent astringent gargle is prepared for mouth irritations. The flesh, either fresh or boiled to a paste, can be added to meat stews for a fruity taste. Dried seeds are used on salads or combined in stuffings. Pomegranate seeds have been utilized for centuries as a cure for tapeworm. Nana is a nearly evergreen shrub to 3 feet tall and grows effectively in a pot. Another larger fruiting variety is Sweet which produces very sweet fruit. Wonderful offers quality fruit almost everywhere except in cool coastal areas.

Culture: All varieties of pomegranate require good drainage and sun. Use equal parts of loam, sand, and peat if planting in a container.

They are not particular about soil, but water deeply and regularly to prevent the fruit from splitting. Let the fruit mature on the tree, or else pick it in the fall and allow it to ripen. Prune to remove suckers, dead wood and hindering branches. In humid climates, a heavy fungus may develop. **Propagate** by suckers, or by using 10- to 12-inch-long hardwood cuttings taken in the dormant season.

POND LILY, *Nymphaea odorata*

Origin: North America.

The white pond lily is found in the still shallow waters of lakes and streams almost everywhere in the United States. Its thick reniform leaves float above the water on long stalks which are attached to a rootstock buried in mud. From June to September, these plants produce spectacular and fragrant flowers that open in the morning and close in the late afternoon. Later, egg-shaped fruits form; full of seeds, they become leathery and podlike. Rootstocks are best transplanted in early spring into waterproof boxes or tubs, and covered with at least 2 feet of water. The thick and starchy rootstocks, gathered after the stalks wither, are a tasty substitute for potatoes. Boiled, baked or fried they have a pleasantly sweet taste; and they may be sliced and added to soups or stews; or sliced, dried, and ground into meal. A tea made from the rootstocks has been used as a gargle and as an eyewash. Rootstocks and leaves are useful as poultices to ease pain and reduce swellings. The young, unfurled leaves may be boiled lightly in salted water and adorned with vinegar dressing. Unopened flower buds are delightful cooked very lightly in boiling water, then topped with butter. Kernels of the large dried seeds have a sweet nutty flavor. They are good boiled and served with butter, or ground into flour. Yellow water lily (*Nuphar adventum*) is similar to the white pond lily and is used for the same purposes.

Culture: Water lilies need four parts of good garden loam enriched with one part of well-rotted cow manure. Give them full sun and feed once a year by burying a handful of blood meal enclosed in a cloth bag into the soil by the roots. If crowns are apt to freeze, cover the container with plywood or store in a basement. **Propagate** by dividing the rootstocks into 6- to 8-inch sections, so each piece has one or two sprouts. Each sprout makes a new lily. Divide every third year.

PRICKLY PEAR, Tuna, *Opuntia megacantha*

Origin: Mexico.

There are many species of *Opuntia* spread across the Americas, surviving such unfavorable conditions as drought, heat, and even long rigorous winters. In the days of the Spaniards, tuna (*O. megacantha*) was introduced into California and has been cultivated ever since for food or decoration. This is a big treelike cactus to 15 feet tall, with a woody trunk and flat, green, jointed pads 15 to 20 inches long. Tiny, curved bristles on the fleshy pads pierce the skin upon contact. Large glamorous orange or yellow flowers bloom in spring or early summer, turning to red pear-shaped fruits which are sweet and good-tasting. These fruits may be eaten raw. Just slice off the ends, slit the skin lengthwise, remove the small seeds, and scoop out the gelatinous pulp. The hard seeds give variety to soups, or may be ground into meal. Raw pulp is a remedy for sunburn. It is delicious added to salads or made into syrup. The syrup base is great to make candy, jams, and jellies. Dried in the sun, the pulp is prepared for the table like any dried fruit. After the skin and bristles are removed, young tuna pads are roasted or cut up in strips to be boiled like a vegetable. They are also excellent pickled. A peeled older pad soaked in water can be applied as a poultice on bruises and wounds to deaden pain and promote healing. For pulmonary problems, a boiled drink is made from the juicy pads. Flower buds are tasty, either raw or cooked.

Culture: Prickly pears need well-drained sandy soil, full sun, and a chance to dry out between waterings. Reduce waterings in the fall to let the plants go dormant. Feed monthly in spring and summer with an all-purpose fertilizer. These plants are subject to scale insects and mealy bugs. Root rot occurs from overwatering. **Propagate** by breaking or cutting off pads, drying them off a few days, then rooting them in sand.

QUINCE, *Cydonia oblonga*

Origin: Persia and Turkestan.

Long before the Christian era, the quince was cultivated for its lumpy, hard, golden-yellow fruit. The deciduous shrub or small tree grows slowly to 25 feet tall and thrives anywhere apple trees do. Pruned into tree form or left as a shrub, quince remains attractive all year. Fruits set from the dainty white (sometimes light pink) flowers and become ripe anywhere from late September to October. The plump fruits require gentle handling to prevent bruising. For best results, pick sound fruit and store it in a cool, frostproof place to continue ripening. Classic varieties are Apple, Champion, Pineapple and Smyrna. Each has a delicate, sweet fragrance of its own. Flesh of the quince remains woody in texture and puckery to taste, until cooked. Prized because of its high pectin content, it is also a good source for calcium and vitamins A and C. Quince jam, jelly, butter and preserves are tasty, as are conserves and marmalades. Quince is mouth-watering served as a spiced sauce, combined with apples in an open pie, or included in meat loaf. Volumes of seeds boiled in water make a healing mouth wash, or a soothing eye application. A hair setting and skin lotion can also be made from the seeds. *Chaenomeles*, known as flowering quince, comes from eastern Asia. This popular garden shrub and its many varieties is one of the first to bloom in the spring. Masses of spectacular flowers cover the branches before the shiny green leaves appear. Most varieties are thorny. Some bear small quince fruits which may be used the same way as the common quince.

Culture: Quinces tolerate light to heavy soil if it is well-drained. They like sun and a thick straw mulch. On fertile soils, fertilizer is unnecessary. Prune to thin out a heavy fruit crop, and to stimulate growth. Control borers and fireblight. **Propagate** by layering, mounding, suckers or hardwood cuttings.

RHUBARB, *Rheum rhaponticum*

Origin: Asia.

This exotic vegetable will produce its palatable bright red or green leaf stalks in most parts of the United States, providing the right variety is selected for the region. At least three or four clumps should be planted per family, with each planting spaced 2 to 4 feet apart. For color in the perennial garden, nothing can compete with these striking leaf stalks and their large, red-veined leaves. The poisonous, dark green leaves rise from a big, fleshy rootstock which penetrates below ground. The inside of the rootstock can be chopped or ground into a powder and used as an astringent for diarrhea. At the end of a central stalk, small, greenish-white flowers often form in dense panicled clusters. Blossom stalks should always be removed, but the flowers may be fried in butter and dusted with powdered sugar for a tasty new food. Since rhubarb seldom comes true from seed, it is best to have a seedless variety such as MacDonald. Two years must elapse before the first pulling of these succulent stalks for eating. They are an excellent source of vitamin A and contain considerable calcium. As a laxative, they are safe and dependable cooked. Rhubarb stalks are very tasty used as a fruit in pies, sauces, tarts, jams, jellies and marmalades.

Culture: Rhubarb needs deep, rich, well-drained soil. Provide plants with some shade where summers are hot. Never let the roots dry out. Water deeply when the tops show active growth. Cover the roots with

a thick layer of cow manure or compost each fall and, in the spring, dig it into the soil. Pick off beetles found boring into crowns, roots, stalks and stems. Destroy plants that develop root rot. **Propagate** by dividing crowns with roots attached, with each division containing one or more buds. Plant in early spring or late winter, setting the bud tops 4 inches deep. Rhubarb can remain undisturbed for about ten years.

ROSE, *Rosa spp.*

Origin: North Temperate Zone.

No garden is complete without roses and none are more lovely than our own native wild roses that form thickets along fences, open woods, stream banks or meadows. Wild roses are very hardy, tolerating cold, heat, dryness and salty air. Vigorous growers, they make excellent hedges and borders for gardens. The flowers, usually blooming in June, contain a spicy fragrance and produce masses of color. Of the highest worth of escaped roses, *R. rugosa*, with red petals, achieves the most intensity and strongest flavor. The acids built in the petals are said to be helpful in desolving gallstones. Dried rose petal tea is used in folk medicine for dizziness, headaches, and as a heart tonic. The tea is also employed as an eyewash for conjunctivitis; and mixed with honey, it is useful as a gargle for swollen gums. Rose petals are eaten raw in salads, made into syrup, jams, jellies, or candied. Developing from the flowers are the predominately red berries called hips. Rose hips generate more vitamin C than orange juice and will hang on the bush the entire year. Eating them daily promotes good health. Try them raw, whole, chopped into salads, or stewed. Popular rose hip tea is made by finely chopping the hips. Hips are also beneficial processed into a purée, jam or jelly. Pulverized into a fine powder, they become a flavoring for food. The seeds, a rich source of vitamin E, may be boiled into a fluid to utilize in recipes specifying water—e.g., for jam, jelly or syrup. Dried leaves and roots make an herbal substitute for oriental tea.

Culture: All roses require well-drained soil enriched with organic matter. To perform well, they need full sun and plenty of water. Apply a thick mulch each spring, and encourage new wood through spring prunings. Main insect pests are aphids, spider mites and thrips. Roses are subject to many diseases, including mildew, leaf spot and rust. **Propagate** by softwood or semihardwood cuttings containing four growths of eyes on the stems. Remove all but the two top sets of leaves. Place several cuttings in a covered propagation box. Try one under an inverted glass jar.

ROSEMARY, *Rosemarinus officinalis*

Origin: Mediterranean Region.

In its native land, rugged and picturesque rosemary grows on dry, rocky hillsides near the sea. An evergreen shrub with spiky leaves, it becomes a haze of lavender-blue flowers in spring and sometimes at other seasons. On a hot day, this aromatic herb has the delightful odor of pine needles. It discourages insect pests in the garden, but does attract bees. Although rosemary grows up to 6 feet, there are smaller varieties suitable for groundcovers, hanging baskets and pots. In most parts of America, the rather tender plant must be cared for indoors during winter. A sprig of rosemary and its flowers gives a pungent taste to lamb, chicken, rabbit, or in stews, soups, and sauces. The flavor of such vegetables as broccoli, cauliflower and spinach is improved by a sprinkling of rosemary after cooking. A light touch of rosemary in fruit drinks, fruit salads, omelets and stuffings adds a spicy taste. A fine jelly can be made from this herb. Dried leaves and flowers may be rubbed down and stored in airtight containers. A fragrant nerve tea is made from them. Fresh rosemary tossed into a steaming tub of bath water offers a fresh woodland aroma. Leaves, alone or mixed with camomile, make a good hair rinse. A sprig or rosemary placed in boiling water acts as an inexpensive and pleasant room

freshener. Oil extracted from the plant is used in medicines and perfumes.

Culture: Rosemary needs full sun and poor, dry soil to produce plenty of fragrance. Pick leaves at any time for use. Harvest for drying by cutting off one half of the current season's growth after flowering. This plant resists insects and diseases. **Propagate** by taking stem cuttings (with heels attached) in midsummer and rooting them in sand. Or, use rooted parts of an existing plant.

SAGE, Garden Sage, *Salvia officinalis*

Origin: northern Mediterranean coasts.

Garden sage is common as a wild herb on dry hillsides in its homeland and has been popular since antiquity for culinary and medical purposes. Cultivated for its pungent and aromatic leaves, these hardy, woody, perennials reach 2 feet in height. They can be clipped to form a low hedge, or interspersed in the garden with members of the cabbage family to fight the cabbage moth. The grey-green, pebble-textured foliage is almost evergreen. Slender purple spikes of blossoms peek from dense leafage to entice bees in early summer. Fresh sage leaves contain a rather minty flavor which is strongest after flowering. They find uses in vegetable soups, onion stuffings, sausages, all roasts, rabbit, fish and poultry dishes. Chopped fresh leaves blended with mustard and rubbed over meat not only add flavor, but the mustard tenderizes the meat. Sage becomes more palatable in teas when lemon juice is added. Teas made from the fresh or dried leaves are useful as a gargle for mouth sores, laryngitis and inflamed throat. The brewed tea is good as a spring tonic, for headache, liver and kidney disorders, to soothe nerves and reduce fever. Fresh leaves rubbed on gums strengthen and harden them. Combined with equal parts of oriental tea, they darken the hair. A tea made from the dried leaves is claimed to restore hair and remedy dandruff. The leaves of sage make an easily digested jelly. Dried leaves and seed pods provide adornment for inside the house.

Culture: Sage requires a good well-drained garden soil, full sun, and should be kept on the dry side. In the fall, cut back the plant to 8 inches to promote young, tender leaf growth in spring. Destroy caterpillars that strip the leaves. **Propagate** by cuttings taken in the fall, by division of rootstocks, or by layering. Replace plants that are three or four years old, as they tend to become too woody and develop fewer leaves.

SASSAFRAS, *Sassafras albidum*

Origin: North America.

This tall native shrub, or usually medium-sized tree, is distributed from New England west to the Mississippi Valley, south to Florida and Texas. It is found in open woods, abandoned fields, along fences and roads, or in other dry places. Sassafras has gray, irregularly ridged bark, with delightfully scented foliage that turns flaming orange and scarlet in the fall. The deciduous leaves vary from oval to mitten-shaped, while the golden-yellow clustered blossoms, appearing in the spring before the leaves unfold, have a spicy smell. In the fall, pea-sized, dark blueish fruits ripen on fleshy red stems. These are borne on the female plants and are relished by birds. Cultivated, sassafras makes a perfect hedge or an excellent shade tree. A fragrant pink tea made from the root bark has been popular for generations as a stimulating and warming tonic. The hot tea was once prescribed to treat high blood pressure and to promote perspiration. Delicious jellies come from strong sassafras tea. The aromatic bark and roots are good to flavor sauces, and can be chewed to relieve toothaches. Containing a mucilaginous substance, the chopped, fresh or dried, leaves are used to thicken and season gumbo, gravies and soups. Dried flowers and stems are made into a tea to reduce fevers; the fruits mixed in wine for colds. Young shoots may be cut in the spring to concoct beer. Rubbed over the skin, sassafras repels insects. It is used in the drug industry to flavor unpleasant tasting medicines.

Culture: Sassafras grows best in deep, well-drained, sandy soil and in a sunny location. It requires watering in arid regions. To produce fruits, two plants are necessary. Remove suckers that form around the plant's base. Prune out branches affected by cankers. **Propagate** by root cuttings or suckers. Due to the long root system, sassafras is difficult to transplant from the wild.

SERVICEBERRY, Juneberry, Shadblow, *Amelanchier spp.*

Origin: North Temperate Zone.

Many species of serviceberry are established in the eastern and western United States, thriving in woods under larger trees or along moist hillsides and stream banks. All are deciduous shrubs or small trees, often with multiple trunks that are ash gray in color. Each presents glowing fall foliage and outstanding branch patterns in winter. The long-lived and highly ornamental serviceberries are not hard to maintain beneath larger plants or in containers. In early spring a profusion of dainty white blossoms opens before the leaves unfurl, replaced in early summer by loose bunches of small, sweet fruits. The bony-seeded fruits may be eaten raw, or dried and used as raisins in recipes. When cooked, the seeds soften and share an almondlike flavor with the fruits, which are used in jam, jelly and purée making. The tasty fruit makes superb pies, or can be saved by canning. Western serviceberry (*A. alnifolia*) grows shrublike to 15 feet and has clusters of small blueish berries. Shadblow (*A. canadensis*) eventually becomes a tree 30 feet tall that displays small maroon-red berries. Its soft grayish leaves are

especially attractive in the spring. Apple serviceberry (*A. grandiflora*) is a natural hybrid with spreading branches, slowly reaching 30 feet in height. The large flowers, resembling apple blossoms, develop into purplish-black fruits. June-berry (*A. stolonifera*) is a 4-foot shrub which spreads by suckers. Its small, purple fruit has a sweet taste.

Culture: Serviceberries do well in light shade and moist, well-drained soil. They rarely need pruning or fertilizing. Fireblight and spider mites may be troublesome. Yellowing and dropping of spring leaves indicates juniper rust. Remove and burn infected plants. **Propagate** by layering or suckers.

SPEEDWELL, *Veronica officinalis (V. serpyllifolia)*

Origin: North America.

There are numerous wild speedwells, but this one is a semi-prostrate perennial herb which is widespread in the United States. It pops up in lawns, meadows, fields and woods, creeping and rooting readily at its joints. Each woody, hairy stem sends up branches to 10 inches high and forms a dense mat at the base. Leaves are grayish-green, soft and smooth. From May to August, pale blue flowers with darker stripes grow in clusters at the top of the stems. The capsule fruit contains a large number of seeds. Although usually regarded as a bothersome weed, speedwell can be attractive either spilling over a container, or protruding from a rock crevice. When picked in late spring and summer, the succulent and fairly pungent leaves make an excellent substitute for watercress. Chopped leaves contribute to all types of salads, being

particularly delectable with vinegar dressings. Cooked as a vegetable, speedwell is a healthful food high in vitamin C. It is a valuable remedy for coughs accompanied by phlegm. An infusion of the fresh or dried flowering plant may be used for migraine headaches, stomach maladies, or as a gargle for mouth and throat sores. Taken in large quantities, fresh juice from the plant is supposed to be beneficial for gout.

Culture: Speedwell prefers a rich, moist, well-drained soil. It takes full sun or very light shade. Watering is necessary throughout dry spells, especially while blooming. This plant is practically immune to insects and diseases. **Propagate** by division after flowering. Divide plants every third year.

STRAWBERRY, *Fragaria spp.*

Origin: North America.

The vigorous wild strawberry is well-distributed across the United States and grows in patches under forest trees, by shady banks, open fields or clearings. From the Pacific Coast native *F. chiloensis* many cultivated varieties have been derived. All strawberries are accommodating perennials and can be grown in a variety of situations, even substituted for a lawn. If space is limited, they thrive in any kind of container, a vertical garden, or a hanging basket. No cultivated variety contains the same luscious flavor or is as rich in vitamins and calories as the tiny wild strawberry. Eaten in quantity, raw or cooked it has a slightly laxative effect. The juicy berries are splendid with shortcake, and in pies, puddings and jams. A crushed berry rubbed over the face removes a mild sunburn and whitens the skin. Fresh berry juice is useful as a dentifrice. Mashed berries made into a drink quench thirst and reduce fever conditions. Wild strawberry leaves are a rich source of vitamin C and iron. Juice extracted from them produces a fine breakfast beverage. Strawberry-leaf tea offers a bracing tonic for children and convalescents. The tea used both internally and externally has been reported to cure acne and eczema.

Culture: Strawberries like a rich, moisture-retaining soil that is loose and somewhat acidic. An abundance of stable manure, full sun, and quantities of water are needed while the fruits grow. A mulch of bark chips or straw modifies the soil temperature and keeps the fruit clean. Destroy aphids, leafhoppers or mites if they attack. Remove and burn plants affected with leaf spot or leaf scorch. **Propagate** by cutting off runners. Runnerless varieties must be propagated by dividing the crowns into several pieces. Set the plants so the buds are level with the surface of the soil.

VIOLET, *Viola spp.*

Origin: North Temperate Zone.

At least thirty species of wild violets flourish in the United States, and are commonly found growing in damp meadows, fields, and along edges of woods among other low-growing plants. Leaf shapes and sizes of these rather hardy perennials vary considerably. The lovely spring blooming flowers range in color from shades of purple, pink, white and yellow. When cultivated in the garden, violets highlight borders, edgings, or containers, and they also serve as ornamental groundcovers. All kinds are edible, even garden varieties. High in vitamins A and C and iron, a few violet leaves eaten daily provide the daily requirements for these vitamins. Finely-minced young leaves can be used in salads, puddings and jellies; or the leaves and stems can be cooked like spinach. Violet leaves are useful for thickening soups. They can be made into compresses, poultices, and ointments. Raw buds and flowers are good in salads and aspics. Delicious uncooked violet jam retains the complete vitamin content of the flowers as well as their color and flavor. A syrup made from violets is said to be excellent for bronchitis, colds, coughs and asthma. Candied flowers make pretty decorations for cakes and other desserts. Violet tea, made from the dried leaves and flowers, is a health-

ful substitute for oriental tea. The dried plant pounded into a powder and combined with honey cures skin diseases, among them impetigo and scabies.

Culture: Violets do best in moist soil provided with leaf mold. Plant them in full shade where summers are hot. Cut back rank growth in late fall for better spring flowers. For heavy bloom, feed before flowering with a 5-10-5 fertilizer. Violets are attacked by red spiders. Most common diseases are leaf spot and root rot. **Propagate** by runners taken in early fall.

WATER CHESTNUT, Water Caltrop, *Trapa natans*

Origin: Eurasia.

The water chestnut was introduced into North America by early settlers and has become widely naturalized in quiet ponds and streams throughout the eastern United States. It is cultivated for its savory fruits and as an ornament to aquariums, pools, and waterproof containers. This aquatic herb produces beautiful, floating clusters of leaf rosettes, often exquisitely mottled or variegated. Small white flowers bloom in summer, borne on the surface leaves. Eventually the leafstalks dip into the water and stay in the muck while the fruits ripen. By winter, the leaves have died back to the rootstock and the fruits have broken away from the parent plant. Then they are ready for harvesting. The 4-pronged, fleshy chestnuts grow to 2 inches across and have a nutlike flavor when peeled. Rich in carbohydrates, they are nourishing raw or roasted. They can be used thinly sliced as carriers of cheese mixtures, on salads, or in hors d'oeuvres. In oriental recipes they add a crisp, nutty taste. The nuts are splendid boiled for five minutes with cabbage and sliced celery. Water chestnuts can be ground into meal, or candied in the same way as true chestnuts. *T. bicornis*, known as Ling Ko, is somewhat similar except that it has two prongs, instead of four, to the fruits. In either species, thirty to forty planted chestnuts should yield a crop of one hundred nuts.

Culture: Water chestnuts require the same treatment as the pond lily (see page 123). **Propagate** by replanting part of the second year's harvest. Nuts for planting are available from nurseries specializing in water plants.

WILD CHERRY, Wild Plum, *Prunus spp.*

Origin: North Temperate Zone.

Various kinds of wild cherry and wild plum are widely distributed across the United States, occurring near banks of streams, fences and roadsides. Members of this group are closely related and often confused because of their similarity. In general, wild cherry has fruit bunches a little shorter and the usually smaller fruit ripens a little later than wild plum. All are deciduous shrubs or small trees, producing either sweet or bittersweet edible fruits which vary in color from red through purple to blueish to almost black. Occasionally yellow-fruited plants appear. In springtime, fragrant clusters of white flowers precede the ripening fruits of late summer. Dried wild cherry bark, prepared as a tea, was once a popular remedy for colds, coughs, and diarrhea. The dried, boiled and scraped inner bark of wild plum was made into a solution for treating mouth ulcers. Using quantities of the bitter barks can cause poisoning and therefore should be prescribed by a doctor. Although sometimes quite sour, the wild fruits are edible raw. All kinds make a tasty jelly when combined with apple juice. The cooked juice with honey added is a good punch, or a syrup on ice cream, puddings, and pancakes. Plum butter, jam and preserves have a tangy taste. Pitted wild cherries and plums make flavorful pies, sauces and tarts.

Whether trained as trees or clipped as hedges, most varieties of these hardy natives are desirable in a garden and grow anywhere apples will grow.

Culture: Wild cherry and wild plum do well in almost any soil if they have full sun or light shade and a normal amount of watering. Once established, they become quite drought tolerant. Be alert for attacks from aphids, leaf miners, tent caterpillars and whiteflies. **Propagate** by layering, suckers, or softwood cuttings placed in a covered propagation box.

WILD LILAC, New Jersey Tea, *Ceanothus spp.*

Origin: North America.

These mostly evergreen shrubs, small trees or low groundcovers, favor hillsides near dry, open woods. *Ceanothus* are found in the eastern United States, the Intermountain Region, and the Northwest, but the greatest number are native to California. Although considered rather short-lived, they compensate by their hardiness and easy to cultivate qualities if grown on mounded soil and if water is kept away from the crowns. Leaves of *Ceanothus* vary from big, tiny, oval or even hollylike, but are nearly always 3-ribbed or strongly-veined at the base. A tea is made from the leaves which are best picked while the plant is in bloom. The leaves combined with herbs such as bee balm, camomile, or mint, produce a tastier tea. New Jersey tea was made from the leaves of *C. americanus* during the American Revolution and was one of the main substitutes for imported tea. Lovely spring flowers burst open in dense spikes or clusters that attract bees. They range in color from pure white through all shades of blue to almost purple. The astringent properties of the root bark resulted in its use as a medicine. Dried powdered bark was once applied as a cure for venereal sores. The dried and ground bark was made into a tea to correct stomach and bowel disorders, and to help asthma and bronchitis. The tea is useful as a gargle for sores in the mouth and swollen tonsils. An infusion of the bark becomes a good tonic when served hot.

Culture: Wild lilac prefers a light, porous soil and open sunlight. Good drainage is essential. Mulch with leaf mold, but give little or no fertilizer. Allow the soil to dry out between waterings, then apply water deeply. Prune back the stems during the growing season to shape and control growth. These plants are subject to aphids, scale insects and whiteflies. **Propagate** by hardwood or softwood cuttings, or by layering.

YUCCA, *Yucca spp.*

Origin: North America.

Several species of yucca grow abundantly in the southwestern United States, appearing in deserts, on dry plains, and dry mountain slopes. Most of these bold and striking evergreen perennials accept garden situations. Growing from 2 to 20 feet or more, they look well grouped in desert gardens or among foliage plants with softer leaves. Generally, yuccas have a basal rosette of sword-shaped, leathery leaves, but some culminate into trunked trees. Masses of bell-shaped, greenish-white to cream-colored flowers bloom luxuriantly on thick supporting stocks. Buds and flowers of all yuccas are edible raw, dried, or cooked. They become a delicacy in salads, or boiled lightly. Roots of various species may be boiled for a warm poultice to relieve painful rheumatic joints. Fleshy fruited kinds, such as datil yucca (*Y. baccata*) provide a sweet and nutritious food at maturity, but should be eaten sparingly due to the strong laxative effect. This fruit pulp is served raw or boiled. Partly mature fruits can be picked and ripened indoors. Seed pods are appetizing cooked, and the large black seeds may be eaten roasted or ground into flour. Crushed roots make a creditable soap substitute, particularly good for shampooing hair. Soapweed (*Y. glauca*) is used for the same purposes. Its pleasing fruits have the flavor of squash, either baked or boiled. The flower stalks, cut before the buds open, are sliced into sections and then baked or boiled. Joshua tree (*Y. brevifolia*) grows

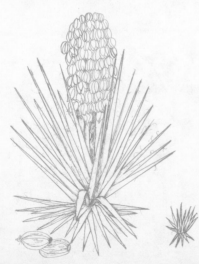

slowly to a tree with a heavy trunk, producing only a few grotesque branches. From February to April it offers a magnificient display of flowers.

Culture: Yuccas thrive in light or sandy, well-drained soil and full sun. They are best suited for hot climates, but will grow in colder areas. Once established, they are drought-tolerant. Crown and root rot diseases develop from overwatering. **Propagate** by offsets.

RETAIL NURSERY SOURCES FOR HARD-TO-FIND-PLANTS

The purpose of the following list of mail-order firms, though by no means complete, is to ease the search for hard-to-find plants and/or seeds. Catalogs are free of charge unless otherwise indicated.

American Perennial Gardens, 6975 Dover Street, Garden City, Mich. 48135. Specialize in nothing but good perennials. Catalog $ .25.

Bergen's Flower Center, 622 Vernon Ave. East, Fergus Falls, Minnesota 56537. Uncommon plants suitable for cold climates. Several garden centers in Minnesota. Price list for particular items.

Burgess, Box 348, Galesburg, Mich. 49053. A good selection of dwarf citrus trees. Two-year-old plants often shipped with small fruit growing on them. Popular foliage and flowering plants. Mail-order catalog.

Cook's Geranium Nursery, 712 North Grand, Lyons, Kansas 67554. New, old, rare and imported varieties of geraniums. Catalog $.35.

Dean Foster Nurseries, Hartford, Michigan 49057. Stocks all kinds of berry, fruit and nut trees; also American grape varieties, ginseng and vegetable roots. Mail-order catalog.

Gardens of the Blue Ridge, P. O. Box 10, Pineola, North Carolina 28662. Native trees and shrubs, wide selection of wildflowers. Mail-order catalog.

Henrietta's Nursery, 1345 N. Brawley Ave., Fresno, California 93711. Rare species of cacti and succulent plants. Mail-order catalog $.35.

Jackson & Perkins Co., Medford, Oregon 97501. Lists dwarf fruit trees, grapes, asparagus, rhubarb and berry plants in their free catalog. The latest in prize-winning roses.

K & L Cactus Nursery, 12712 Stockton Blvd., Galt, CA 95632. A wide assortment of cacti and succulents; some good bargains for beginners. Beautifully illustrated catalog $1.00.

Lamb Nurseries, East 101 Sharp Ave., Spokane, Washington 99202. Dwarf shrubs, rock garden plants, perennials, groundcovers, herbs, hardy succulents. Mail-order catalog.

Lounsberry Gardens, Box 135, Oakford, Illinois 62673. Hardy wild-flowers, ferns, perennials and rock garden plants. Catalog $.25.

Mellinger's, Inc., North Lima 18, Ohio 44452. An extensive variety of trees and shrubs in small sizes, many unusual, including ginseng and some tree seeds. They also sell ladybugs and preying mantis eggs. Mail-order catalog.

Nichols Garden Nursery, 1190 N. Pacific Highway, Albany, Oregon 93721. Features herbs, rare plants and organic seeds. Mail-order catalog.

George W. Park Seed Co., Inc., P. O. Box 31, Greenwood, South Carolina 29647. Large supplier of African violet, cactus, flower, herb and vegetable seeds. Some unusual perennial plants. Mail-order catalog.

Clyde Robin, Box 2855, Castro Valley, California 94546. More than 1200 different kinds of seeds for native plants, trees, and shrubs. Mail-order catalog $1.00.

The Shop in the Sierra, Box 1-N, Midpines, California 95345. Western native trees, shrubs, and groundcovers. Catalog $1.00.

Roses of Yesterday and Today, 802 Brown's Valley Rd., Watsonville, California 95076. Wide selection of old-fashioned and modern roses. Catalog is $1.00 and two free catalogs are sent to customer-supplied addresses for an order totaling $10.00 or more.

William Tricker, Inc., 74 Allendale Ave., Saddle River, New Jersey 07458; also, 7125 Tanglewood Dr., Independence, Ohio 44131. Water lilies and bog plants. Mail-order catalog.

Valley Nursery, Box 4845, 2801 N. Montana, Helena, Montana 59601. Uncommon trees and shrubs for cold climates. Seedlings and transplants. Price list available.

Van Ness Water Gardens, 2460 N. Euclid Ave., Upland, California 91786. Water chestnuts, water lilies, lotus and other water plants for growing in containers. Catalog $.50.

The Wild Garden, George Schenk, Box 487, Bothell, Washington 98011. Large variety of hard-to-find rock garden plants. Ericaceous plants. Catalog $1.00, deductible on first purchase.

SOURCES OF INFORMATION

For those readers who wish to make an in-depth study of plant care and uses, this list of reference books and pamphlets will prove to be beneficial.

Campbell, Mary Mason, *Betty Crocker's Kitchen Gardens*, Scribners, N. Y., 1974. Promotes ideas for uses of herbs and other garden plants.

Coon, Nelson, *Using Plants for Healing*, Hearthside Press, Inc., Great Neck, N. Y., 1963. An excellent book, devoting a page to each of about 150 different plants which American herb doctors have used.

Furlong, Marjorie & Pill, Virginia, *Wild Edible Fruits and Berries*, Naturegraph Publishers, Happy Camp, CA, 1974. Includes locations of plants and delicious recipes.

Hansen, C. J. & Hartmann, H. T., *Propagation of Temperate-zone Fruit Plants*, Circular 471 revised, Div. of Agricultural Science, University of California Press, Berkeley, 1962. This pamphlet also explains the principles of budding and grafting.

Kirk, Donald R., *Wild Edible Plants of the Western United States*, Naturegraph Publishers, Happy Camp, CA, 1970. Covers nearly 2000 useful plants, telling how to identify and prepare them.

Lust, Dr. John, *The Herb Book*, Benedict Publications, N. Y., 1974. An M. D. offers his methods and prescriptions for healing with herbs. Also practical for locating and identifying herbs in the field.

Mathews, F. Schuyler, *Field Book of American Wild Flowers*, G. P. Putnam's Sons, N. Y., 1955. Fine for location and identification.

Pendergast, Chuck, *Organic Gardening*, Nash Publishing, Los Angeles, CA, 1971. A guide to growing fruits and vegetables in the safe, healthy and natural way.

Rodale Editors, *Organic Plant Protection*, Rodale Books, Emmaus, PA, revised 1976. Gives valuable help in identifying harmful pests and diseases, followed by explicit instructions on how to get rid of them on berries, fruit trees and vegetables.

Rodale Editors, *The Rodale Herb Book*, Rodale Books, Emmaus, PA, 1974. Not only presents the art of growing and using herbs, but tells how to use plants to control pests in the garden.

Scully, Virginia, *A Treasury of American Indian Herbs*, Crown Publishers, N. Y., 1970. Describes how the Indians, particularly of the Rocky Mountain Region, put herbs to use.

Sherman, Dr. Ingrid, *Natural Remedies for Better Health*, Naturegraph Publishers, Happy Camp, CA, 1970. Contains many old household recipes which are easy to prepare from food plants.

Sweet, Muriel, *Common Edible and Useful Plants of the East and Midwest*, Naturegraph Publishers, Happy Camp, CA, 1975. Explains various uses of plants as discovered by Indians, pioneers, and colonial Americans.

Sweet, Muriel, *Common Edible and Useful Plants of the West*, Naturegraph Publishers, Happy Camp, CA, revised 1976. Shows ways to bring back or preserve health by using wild plants of the West.

U. S. Dept. of Agriculture, *Minigardens for Vegetables*, Home and Garden Bulletin #163, Supt. of Documents, U. S. Government Printing Office, Washington D. C. 20402. Gives instructions for growing vegetables in containers.

U. S. Dept. of Agriculture, *U. S. Farmers Bulletin No. 2201*, Supt. of Documents, U. S. Government Printing Office, Washington, D. C. 20402. Detailed instructions on how to propagate ginseng.

INDEX

In this index common names are given for plant families, species and varieties; scientific genera only are in italics. For plants entered which have similar propagational needs to those plants specifically covered by the text, the reader is referred to "see" the relevant page (also explained in paragraph 4 of Introduction).

SOME OTHER BOOKS BY NATUREGRAPH

KNOW YOUR POISONOUS PLANTS, *by Wilma Roberts James, illustrated by Arla Lippsmeyer.* 65 species are described and illustrated on a full page each. 75 additional plants receive a short paragraph plus a small drawing. Particularly helpful to parents, this book alerts all readers to the fact that many common plants are poisonous. In color.

WILD EDIBLE PLANTS OF THE WESTERN UNITED STATES—Color Edition, *by Donald Kirk, illustrated by Janice Kirk.* 16 color plates by Donald Kirk. This popular handbook covers nearly 2000 species of useful plants found in the western United States, southwestern Canada and northwestern Mexico. Over 400 drawings included in 343 pages.

WILD EDIBLE FRUITS AND BERRIES, *by Marjorie Furlong and Virginia Pill.* For each of the 42 wild edible fruits and berries covered, there is a full-color photograph and description. Locations are given and conservation principles encouraged. A recipe section also adds to the reader's enjoyment. This book is useful throughout the United States and Canada.

COMMON EDIBLE AND USEFUL PLANTS OF THE WEST—Revised Edition, *by Muriel Sweet.* 64 pages, 116 illustrations. Learn how Indians and pioneers used plants for food, medicine, shelter, etc. This book has a helpful cross-reference index.

COMMON EDIBLE PLANTS OF THE EAST AND MIDWEST, *by Muriel Sweet.* 80 pages, 69 illustrations. Especially for the eastern half of the United States, this book is similar to its western counterpart. Common edible plants are identified, and various uses by Indians and early settlers are given.

NATURAL REMEDIES FOR BETTER HEALTH, *by Dr. Ingrid Sherman.* Find out from this health book how to use natural foods and drinks, physical, mental and spiritual exercises, as well as other wholesome ways, to obtain a balanced life. 128 pages.

Ask for these books at your local bookstore, or order from Naturegraph Books, Happy Camp, CA 96039. Catalogs are free upon request.